Late Quaternary Palaeoenvironments of Southern Africa

Late Quaternary Palaeoenvironments of Southern Africa

JANETTE DEACON
Department of Archaeology, University of Stellenbosch

and

N. LANCASTER
Department of Geology, Arizona State University

CLARENDON PRESS · OXFORD
1988

Oxford University Press, Walton Street, Oxford OX2 6DP
Oxford New York Toronto
Delhi Bombay Calcutta Madras Karachi
Petaling Jaya Singapore Hong Kong Tokyo
Nairobi Dar es Salaam Cape Town
Melbourne Auckland
and associated companies in
Beirut Berlin Ibadan Nicosia

Oxford is a trade mark of Oxford University Press

Published in the United States
by Oxford University Press, New York

British Library Cataloguing in Publication Data
Deacon, Janette
Late Quaternary palaeoenvironments of Southern Africa.
1. Geology, Stratigraphic—Quaternary
2. Geology—Africa, Southern
3. Paleoecology—Africa, Southern
I. Title II. Lancaster, N.
551.7′9′0968 QE96
ISBN 0-19-854449-9

Library of Congress Cataloging in Publication Data
Data available

Set by Cotswold Typesetting Ltd, Cheltenham
Printed in Great Britain
at the Alden Press, Oxford

Preface

This book began as a joint report presented in March 1984 to the National Programme for Weather, Climate, and Atmosphere Research funded by the Council for Scientific and Industrial Research Cooperative Scientific Programmes (now the Foundation for Research Development) in Pretoria. The report was a literature survey and synthesis of the evidence for Late Quaternary climatic change in Southern Africa with NL reviewing the geomorphic and sedimentary evidence, and JD the biological evidence. It has been up-dated and substantially re-written as a sourcebook and general summary of the state-of-the-art in the study of Late Quaternary palaeoenvironments. The emphases mirror our own interests, namely the geomorphology of desert environments and the archaeology of Stone Age communities in Southern Africa, but we have drawn also on the work of colleagues in other fields.

We are very grateful for the financial support of the CSIR and for the constant encouragement of Dr Wynand Louw, the Coordinator of the Atmosphere Programme. We should also like to acknowledge with grateful thanks the substantial contribution of Professor H. J. Deacon who has been the project supervisor during 1986. He has given considerable stimulus and generous access to unpublished data. He contributed the section on biogeographic implications in chapter 8.2 as well as data from Boomplaas and Klasies River, all of which form part of his highly successful research projects aimed at correlating palaeoenvironmental evidence from a number of different lines of evidence in the southern Cape.

At all times we have received encouragement, critical comments and valuable reviews of sections of this book from Dr John C. Vogel of the National Physical Research Laboratory at the CSIR. We and many others owe him a debt of gratitude for his stimulating contribution to Quaternary research in Southern African and for the large number of radiocarbon dates processed by his laboratory, without which it would be impossible to establish a reliable chronology of palaeoclimatic events in the subcontinent. Together with his colleagues, T. H. E. Heaton and A. S. Talma, Dr Vogel has given permission for the re-drawing of Figures 3.3, and 6.9–6.13.

Permission for figures to be re-drawn is gratefully acknowledged from *Quaternary Research* (Figs 1.3 and 5.8), Cambridge University Press (Fig. 2.3), Chicago University Press (Fig. 2.5), Balkema, Rotterdam (Figs 2.6,

2.8, 3.1, and 5.6), J. F. Thackeray (Fig. 6.7), R. G. Klein (Fig. 6.8), and H. J. Deacon (Fig. 8.1).

The initial synthesis of palaeoclimatic evidence by NL was carried out at the Univesity of Cape Town, where Professor Richard Fuggle, of the Department of Environmental and Geographical Science, gave valuable support and encouragement for the project. Logistic and institutional support for the final phases of NL's work on the book was provided by the Department of Geology, Arizona State University. JD acknowledges with grateful thanks the continued support of the Department of Archaeology at the University of Stellenbosch. The Institute for Cartography at the University of Stellenbosch was responsible for drawing all the figures except 5.2 (J. Lancaster), 6.2 (V. B. Geleijnse), 3.3, 6.3, 6.4, 6.6, 6.7 and 6.9–6.13 (J. Deacon), and 7.1–7.3 (N. Lancaster).

We have enjoyed stimulating discussions on the Late Quaternary environments of Southern Africa with many colleagues, several of whom have read and commented on sections of this book. They include D. M. Avery, P. B. Beaumont, K. W. Butzer, H. J. Deacon, J. R. Grindley, R. G. Klein, R. R. Maud, J. E. Parkington, T. C. Partridge, J. Rogers, U. Rust, A. Scholtz, L. Scott, P. Shaw, A. S. Talma, J. T. Teller, J. F. Thackeray, M. L. Tusenius, P. D. Tyson, E. M. van Zinderen Bakker, J. C. Vogel, and J. D. Ward. Helen Barlow and Judith Lancaster helped considerably with proof-reading and other tasks in the preparation of the manuscript, and we thank them most sincerely.

Finally, we should like to thank our spouses, H. J. Deacon and Judith Lancaster for their assistance and encouragement at all stages of the work. Any flaws or faults can only be our own.

1987

J. D.
N. L.

Contents

1 Introduction 1

1.1 Approaches to Late Quaternary research 1

1.2 Historical review 3

2 Present and past climates of Southern Africa 12

2.1 Present day climates 12

2.2 Palaeoclimatic models 18

3 Sources of palaeoclimatic evidence in Southern Africa 29

3.1 Introduction 29

3.2 Geomorphic and sedimentary sources 29

3.3 Biological and archaeological evidence 36

4 Palaeoenvironments in the desert and semi-desert ecozones 47

4.1 Introduction 47

4.2 The Namib 48

4.3 Chronological summary 56

4.4 The Karoo and Bushmanland 56

4.5 Chronological summary 59

5 Palaeoenvironments in the summer rainfall ecozones 60

5.1 The Kalaharian Ecozone 60

5.2 The northern Kalahari 66

5.3 The southern and south-western Kalahari 69

5.4 The south-eastern Kalahari 73

5.5 Chronological summary for the Kalaharian Ecozone 78

5.6 The Transvaalian Ecozone 79

5.7 Chronological summary for the Transvaalian Ecozone 93

5.8 The Basutolian Ecozone 94

5.9 Chronological summary 108

6 Winter and all-year rainfall regions: the Cape Ecozone 109

6.1 Dune reactivation and stabilization episodes 110

6.2 Evidence for former sea levels in the western and southern Cape 113

6.3 Alluvial deposits of the southern and western Cape 122
6.4 Slope deposits in the southern and western Cape 122
6.5 Coastal caves and cave deposits 123
6.6 Caves and cave deposits in the southern Cape interior 130
6.7 Vegetation changes in the Cape Ecozone 135
6.8 Faunal changes in the Cape Ecozone 138
6.9 Isotopic evidence from the Uitenhage aquifer and Cango speleothem 141
6.10 Archaeological proxy data 144
6.11 Chronological summary 145

7 Summary and conclusions 148

7.1 *c.* 130 000–80 000 BP 148
7.2 *c.* 80 000–50 000 BP 152
7.3 *c.* 50 000–24 000 BP 153
7.4 *c.* 24 000–16 000 BP 155
7.5 *c.* 16 000–10 000 BP 156
7.6 *c.* 10 000–0 BP 157
7.7 Summary of radiocarbon dated primary evidence 161

8 Explaining the palaeoenvironmental patterns 162

8.1 The nature, scale and timing of changes in Late Quaternary palaeoenvironments 162
8.2 Biogeographic implications 168
8.3 Implications for human population distributions 171
8.4 Implications for palaeoclimatic models 174

Appendix: List of sites with radiocarbon dated evidence for palaeoclimatic conditions in Southern Africa 177

References 187

Author index 211

Subject index 217

1 Introduction

1.1 Approaches to Late Quaternary research

The main aim of palaeoclimatic research is to describe variations in climate in the period beyond the range of the meteorological record. The ultimate goal is theoretical: to understand the physical causes of these variations. By applying principles established through modern climatological research, it is possible to understand the mechanisms of change and the pattern of palaeoclimates in time and space. This has considerable practical significance in, for example, studies of the effects of a possible future increase in atmospheric carbon dioxide content. Palaeoclimatic data provide the best means of testing models designed to assess future climatic trends, so there is an important feedback between the present and the past, and between the past and the future. Palaeoclimatic data are also the only means of studying ecosystem history.

Geographically, this book focuses on Southern Africa, i.e. Africa south of the Zambezi River. It therefore includes a wide range of environments and habitats, excluding only tropical forest and glaciated highlands. We assess the current status of Late Quaternary palaeoclimatic studies in this region and set out some of the priorities for future research. In Southern Africa, the Quaternary has been studied for the most part by two distinct groups of researchers: those aligned to the geological sciences who are interested in landforms and sedimentary deposits, and the commercial and non-commercial aspects of soil types and formation; and those who are concerned with archaeology and the prehistory of human settlement in Southern Africa. Both need to discover how old the deposits are, and under what climatic and other environmental conditions they formed.

The Quaternary includes the Pleistocene and Holocene epochs. The 'official' Pliocene–Pleistocene boundary dates to 1.6 Ma at Vrica in Italy (Backman *et al.* 1983) and the Pleistocene–Holocene boundary is placed at 10 000±250 years ago (Fairbridge 1983, p. 238). Within the Pleistocene, the Lower–Middle Pleistocene boundary coincides with the Brunhes–Matuyama palaeomagnetic boundary at 700 000 years ago and the Middle–Upper Pleistocene boundary coincides with the oxygen isotope deep sea core stage 6–5 boundary at about 130–125 000 years ago (Fairbridge 1983; Turon 1984). These dates do not mark periods of abrupt change, but are convenient points in geological time to group together those deposits that contain similar fossils. We shall be focusing on the Late Pleistocene and Holocene, the past 130 000 years or so, with data from both geological and archaeological deposits. We have chosen this time period partly because it is the most securely dated, and partly because it covers a complete cycle from Last Interglacial through Last Glacial to Present Interglacial conditions. The observations give an idea of the scale and timing of climatic changes that have taken place over this time period in a sub-tropical to temperate southern hemisphere continent. Some of the results can be used to infer what climates may have been like in other glacial–interglacial cycles, and what they could be like in the immediate and more distant future.

In palaeoclimatic terms, the Quaternary is one of several periods of earth history that experienced an ice age. In this case, the positioning of the Antarctic continent over the South Pole and other continental movements, including the joining of North and South America, led to the present-day ocean circulation patterns and the development of wind and pressure systems that control our modern climate. Coupled with long-term cyclic changes in the amount of solar radiation received by the earth, these factors combined towards the end of the Pliocene and throughout the Pleistocene to send the earth into a series of glacial and interglacial modes with a periodicity of a little over 100 000 years. No Pleistocene full interglacial has lasted more than 12 000 years and most have spanned about 10 000 years (Fairbridge 1983). Thus, temperatures prevailing worldwide over the past 2 Ma have, on average, been cooler than they are at present and the Present Interglacial can be expected to end within the next 2000 years.

Southern Africa was not glaciated at any time during the Quaternary, but did experience long periods of cooler temperatures (5–9°C cooler than at present) when ice caps expanded at high latitudes and at high altitudes in lower latitudes. There have also been marked changes in rainfall patterns. Both left their mark in the landforms, sediments, and plant and animal fossils, and this is the evidence we shall be using to reconstruct the climates of the past 130 000 years.

Three levels of palaeoclimatic data can be identified: primary data, including sedimentary, floral or faunal analyses that are accurately located in time and space, usually by being within a well-dated sequence of sediments; secondary data, which are climatic conditions inferred from the primary data; and tertiary data comprising higher level palaeoclimatic syntheses based on the primary and secondary data (Hecht *et al.* 1979).

Primary data constitute the building blocks for second-level palaeoclimatic reconstructions so it is essential that the primary sources are assessed carefully before syntheses can be derived from them. There are two problems here: the first is an historical one in that study methods and interpretations have changed over the past century; the second is one of distribution, for primary data are not evenly spread in time and space, nor are the different types of evidence equally important in all regions. For this reason there are regional disparities in the types of palaeoclimatic reconstructions that can be made and this, of course, affects both the secondary and tertiary levels of palaeoclimatic synthesis. In the Kalahari, for example, most of the evidence is for changes in the ratio between precipitation and evaporation derived from lake levels, whereas in the southern Cape climatic parameters have been more easily interpreted from biological remains.

In this review and synthesis of the evidence for Late Quaternary palaeoclimates in Southern Africa, we begin by tracing the historical development of the subject, and then consider the types of primary evidence that can be used as proxy data for the secondary and tertiary levels. We then turn to a review of the evidence for palaeoclimates on a regional basis focusing on the timing and scale of changes that occurred, and finally present a chronological synthesis.

1.2 Historical review

Palaeoclimatic research has undergone some dramatic changes in direction and focus in Southern Africa and an historical review of changing ideas gives an insight into the way in which a discipline develops and grows, warts and all. During the first half of this century an important purpose of studies of past environments was to establish a relative dating scheme. No absolute dating methods were available at the time, and relative dating relied on the laborious mapping of sedimentary sequences and associated fauna and stone artefacts. Climate-stratigraphic schemes assume that major sub-continental climatic changes were local manifestations of the synchronous worldwide Quaternary climatic shifts. Particular types of sediments were interpreted as the result of higher or lower rainfall or temperatures, and changes were fitted into a recognized sequence of palaeoclimatic events. In South Africa, Rogers (1905) was one of the first to draw attention to the possibilities of relating changes in sea-level (and therefore changes in the amount of ice locked up in polar ice caps) to Stone Age artefact assemblages. He showed that Middle Stone Age artefacts overlaid deposits of the +6–8 m beach along the southern Cape coast and therefore post-dated them. He correctly linked the Middle Stone Age with the sea level that dated to the Last Interglacial.

A concerted effort was made in the 1920s and 1930s to establish a palaeoclimatic chronology in Southern Africa based on the pluvial scheme recognized in East Africa (Leakey and Solomon 1929; Van Riet Lowe 1929; Smuts 1932). By the 1950s it had become clear that the theoretical basis for the pluvial theory was faulty and that the field evidence had at times been incorrectly interpreted. Research since the 1960s has concentrated instead on relating evidence for temperature change to the glacial sequence that has been dated best in deep-sea cores. These changing paradigms in palaeoclimatic research have led to a healthy caution amongst researchers and some have stressed that palaeoclimatic reconstruction must be primarily inductive to avoid the temptation to fit observations into a preconceived framework (Zeuner 1952; Flint 1959*a,b*; H. J. Deacon 1983; Butzer 1984*a*). Purely inductive data gathering is an ideal that is not possible to maintain, though, for we inevitably have notions of what we would expect to find. Without them we would not know where to look nor what to study. The saga of the pluvial hypothesis makes palaeoclimatic research in Southern Africa a fascinating cameo of the history of a science and the way in which inductivism and hypothesis testing should and should not be used.

1.2.1 The pluvial theory in East and Southern Africa

As early as the 1890s, Gregory (1921), working on the mapping of glacial features and former high lake levels in the East African rift valley, suggested that these phenomena could be linked to the glacial chronology of western Europe. A detailed study of glacial phenomena in East Africa and Ethiopia was undertaken in 1927 and 1928 by Nilsson (1932, 1940), a member of the Swedish Geological Expedition to East Africa. He tested the hypothesis that Wayland (1934), also a geologist, who had extensive field experience in Uganda, had formally proposed: that periods of higher rainfall, or pluvials, were coincident in Africa with periods of glacial advance in Europe. The assumption was that lower temperatures during glaciations would reduce evapotranspiration in temperate and tropical latitudes resulting in higher effective precipitation. Wayland reasoned that if this were so, he could expect to find field evidence for major pluvials coincident with the glacials of the northern hemisphere. Nilsson indeed found evidence for higher lake levels and more extensive glaciers on the higher mountains and not surprisingly linked them with glacial advances recognized in Europe at that time.

Between 1926 and 1935, further observations were made by Leakey and Solomon (1929) who postulated a sequence of pluvials and two Holocene wet phases in higher lake levels and river terraces in East Africa. Later, Solomon (1939) repudiated the scheme, but Leakey made a formal proposal for the acceptance of a four-phase pluvial chronology to match the four-phase glacial scheme (Günz, Mindel, Riss, and Würm) at the first Pan-African Congress on Prehistory held in Nairobi in 1947. The pluvials were named, from oldest to youngest, Kageran, Kamasian, Kanjeran, and Gamblian, with the Makalian and Nakuran wet phases in the Holocene. Interpluvials or drier phases contemporary with interglacials were numbered from First to Third (Table 1.1).

In Southern Africa during the 1920s and 1930s there seem to have been two schools of thought. Some, like Rogers (1922), supported the ideas of Penck (1905, 1914) that climatic belts had shifted equatorwards during glacials bringing drier conditions to some areas and wetter conditions to others. On the other hand, Leakey and Solomon (1929), Van Riet Lowe (1929), Smuts (1932), and Wayland (1934) applied the East African pluvial sequence further afield in the belief that glacials were wetter worldwide. Krige (1932) gave an interesting, if implausible, suggestion, that the Pleistocene and other ice ages had been triggered by warmer oceans resulting from continental drift. As a continent was displaced, the hot magma of the substratum was exposed to ocean water behind it and this acted as a heat source. Warmer ocean waters in turn promoted evaporation and increased cloudiness leading to a build-up of ice at high latitudes.

The pluvial scheme offered a much-needed relative dating chronology. Detailed geomorphological studies were made of the Vaal River terraces by Sohnge *et al.* (1937) and of the Olifants River terraces by Mabbutt (1957) (Fig. 1.1) in South Africa. Others were undertaken along the Zambezi by Armstrong and Jones (1936), and Clark (1950), in Zimbabwe by Maufe (1929, 1930), and in Namibia by Korn and Martin (1957), and all were fitted into the pluvial scheme. Some measure of relative dating to assist in this interregional cross-correlation was

Table 1.1. Comparison of African pluvial and Alpine glacial terms

Geological time	Glacials	Pluvials/ interpluvials	Wet phases
Holocene			Nakuran Makalian
Upper Pleistocene	Würm	Gamblian Third Interpluvial	
Middle Pleistocene	Riss	Kanjeran Second Interpluvial	
	Mindel	Kamasian First Interpluvial	
Lower Pleistocene	Günz	Kageran	

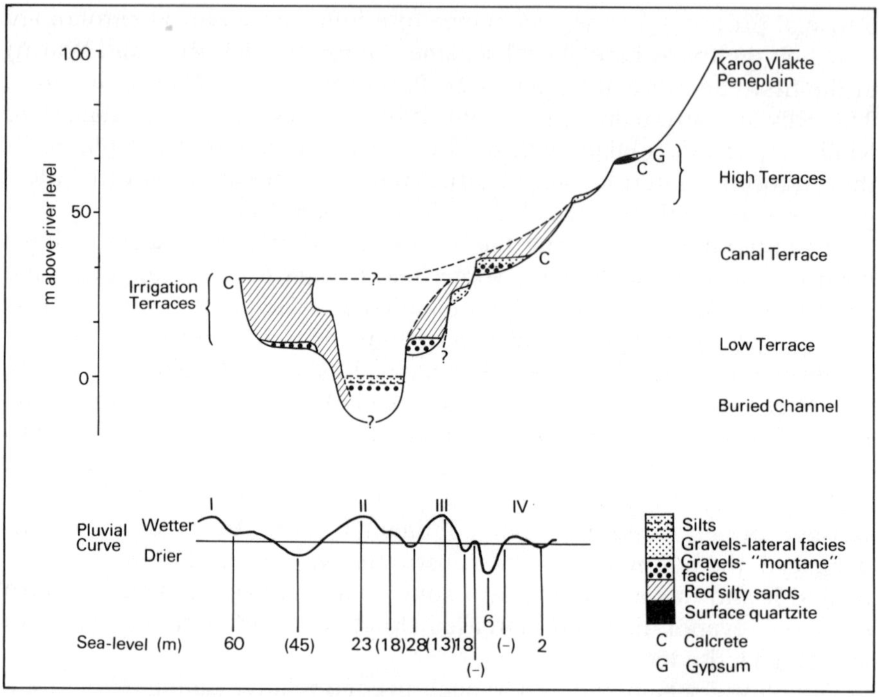

etres

Fig. 1.1 The Olifants River terrace sequence (from Mabbutt 1957).

possible using the associated faunal assemblages and stone artefacts, but by and large the unifying hypothesis was the pluvial theory. The fact that the assumed correlation between glacial and pluvial and interglacial and interpluvial had not been adequately tested did not seem to deter researchers. A cautionary note was sounded at the Third Pan-African Congress on Prehistory in Livingstone in 1955 where two recommendations were made. First, the pluvial terms should be used only in a climatic (and not a relative dating) sense and then only in the type area. Secondly, they should be applied as climate-stratigraphic units outside East Africa only when correlation could be established on at least two lines of evidence (geological, palaeontological, and archaeological). Unfortunately, the terms continued to be used somewhat indiscriminately.

Doubts surfaced again in the late 1950s when Cooke (1957) and Flint (1959*a*, 1959*b*) followed an earlier appraisal of Du Toit (1947) that possible sources of error in the interpretation of the field evidence for pluvials in East Africa needed to be investigated. There was little doubt that cooler periods which resulted in the buildup of ice sheets at higher latitudes during the Pleistocene were the result of world-wide phenomena. It was incorrect to assume, though, that glacial periods were synchronous with wetter conditions in lower latitudes. Nor was it possible to infer a simple relationship between periods of prolonged higher rainfall and higher lake levels, and/or downcutting in river valleys. Flint (1959*a*) and Cooke (1957) recommended that the pluvial scheme as a formal relative dating chronology be abandoned.

From an historical perspective the pluvial hypothesis was something of a red herring demonstrating the danger of setting up a faulty framework and fitting observations into it. As Flint (1959*a*, p. 279) argued

> Our system of Pleistocene geologic stratigraphy and indeed geologic stratigraphy as a whole . . . has been built up on a basis of layers of rock or sediment, defined and described, and correlated from one place to another by means of the fossils they contain . . . So far as I am aware, such merits have not been shown for the climatic scheme. That scheme has not been described nor explained as a system; it seems to have grown little by little, reaching its existing status in default of contrary opinion.

What is perhaps one of the most important points to understand in rejecting climate as the primary basis for a stratigraphic scheme is that climatic concepts are secondary level data, i.e. second-order derivations. They are not stratigraphic entities in the geological sense, but are abstractions derived from the interpretation of strata in terms of the climatic conditions that prevailed during their formation.

1.2.2 Aftermath of the pluvial hypothesis

Palaeoclimatic studies having been linked for so long to relative dating, new perspectives were explored in both theory and fieldwork. Bernard (1962) using astronomical regularities noted by Milankovitch and others, showed that the effects of increased or decreased insolation would have had different effects at different latitudes. His scheme (Fig. 1.2.) illustrates how various rainfall regimes in Africa could have been out of phase. This was the first attempt to set up a detailed hypothesis of rainfall changes in the continent that could be tested in the field, but few researchers paid it much attention. The high lake levels of the Early Holocene in East Africa (Butzer *et al.* 1972; Street and Grove 1976) have shown that at least one of his predictions is correct.

Fascination with the possible effects of increased and decreased rainfall still lingered. Cooke (1962, 1964), for example, used a simplified vegetation map of Africa to estimate the scale of change that may have occurred had rainfall been 50–60 per cent and 140–150 per cent of the present levels at any time in the past. Lee (1963) used the same principal to establish what he termed 'regions of permanent favourability' for hunter-gatherers. Models such as these assumed the same magnitude of

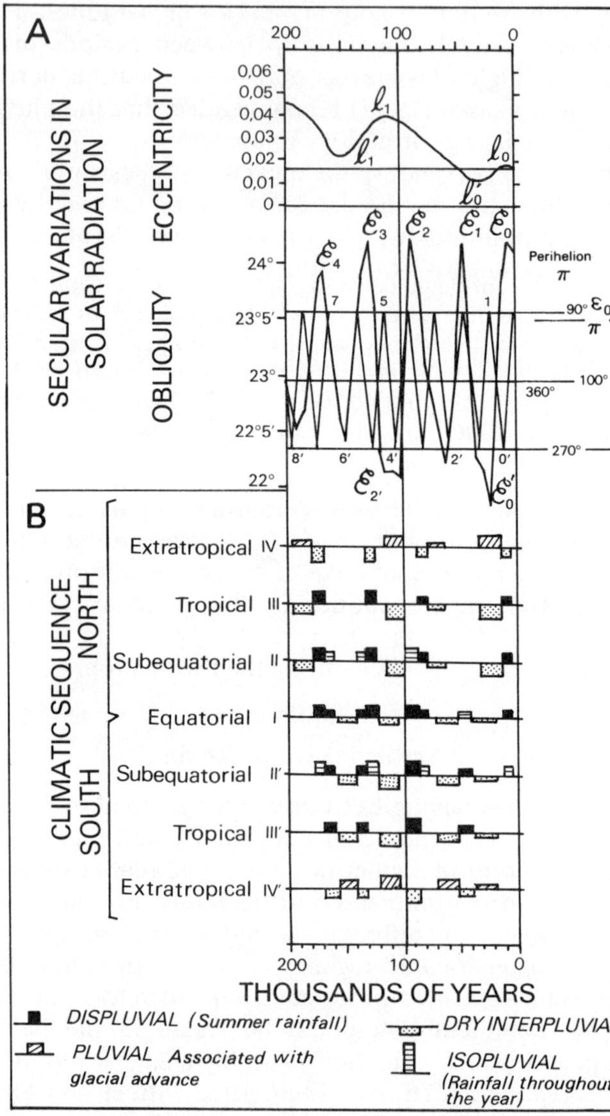

Fig. 1.2 A model of predicted 'pluvials' and 'interpluvials' at different latitudes in Africa over the last 200 000 years based on secular variations in solar radiation (after Bernard 1962).

rainfall changes throughout the sub-continent, not taking note of Bernard's contention that global temperature changes would not have affected precipitation in the same way in all regions.

Brain (1958, 1967, 1969) pioneered the use of sediment analyses to trace rainfall and run-off changes in cave and rock shelter deposits, and the 1960s saw the growth of interest in faunal and floral remains at archaeological sites (Hendey and Singer 1965; H. J. Deacon 1970; Klein 1972*a,b*). These studies were aided of course by the availability of radiocarbon dating which provided a dating method that was far superior to the pluvial scheme, but the most direct palaeoclimatic evidence came from palynological studies by Van Zinderen Bakker (1957, 1962*a,b*, 1967), Martin (1962, 1968), and Coetzee (1967). Van Zinderen Bakker in particular was instrumental in stimulating interest through palaeoclimatic models that initially (1967, fig. 2.5) assumed the equatorward movement of climatic belts. In the 1960s Van Zinderen Bakker boosted palaeoenvironmental research by starting the journal *Palaeoecology of Africa*, and the Southern African Society for Quaternary Research was founded in 1969.

During the 1970s, with encouragement from new techniques for dating and analysis, long-sequence cave sites were excavated, and it became clear that the faunal and floral remains had considerable potential for palaeoenvironmental reconstruction (Klein 1972*a,b*, 1980; Beaumont *et al.* 1978; H. J. Deacon 1979), and Butzer (1973*a*, 1978, 1979) and Tankard (1976*a*) used cave sediments to trace climatic changes. The excavation of Boomplaas Cave in the southern Cape was designed expressly to recover as many different lines of palaeoclimatic evidence as possible from the same stratigraphic sequence and although the project took ten years to complete (H. J. Deacon 1979; H. J. Deacon *et al.* 1984) it has provided one of the most reliable and detailed Late Quaternary climatic records in Southern Africa. Geomorphological studies by Partridge and Brink (1967), Sparrow (1967, 1971), Netterberg (1969, 1978), Butzer and Helgren (1972), H. J. Cooke (1975, 1980), Davies (1976, 1980), Lancaster (1978*a,b*), Helgren (1979), and others explored new avenues for palaeoclimatic reconstruction that included periglacial phenomena, sea levels, calcretes, and pedogenesis.

The model currently dominating palaeoclimatic studies is a modified version of the astronomical theory developed by Milankovitch between 1920 and 1938 and based on variations in the eccentricity of the earth's orbit around the sun, variations in axial tilt and variations in the precession of the equinoxes. It was not until the 1970s that well documented confirmation was found in deep-sea cores for the predicted 41 000- and 23 000-year cycles generated by these phenomena (Imbrie and Imbrie 1979). The timing and regularity of these global changes can be shown from deep-sea cores to have been broadly synchronous throughout the world. However, land-based observations of oxygen isotope changes in the Antarctic ice sheet and of glacial advances and retreats in glaciated southern hemisphere countries suggest that at least at the end of the last glacial cycle, warming of the southern hemisphere preceded that in the northern hemisphere by several thousand years (Salinger 1981, 1984). The importance to Southern Africa of the Late Quaternary palaeoclimatic record of Antarctica was recognized by Van Zinderen Bakker (1978) who organized an international conference on Antarctica and southern hemisphere palaeoclimates in 1977. Research reports were updated at a second meeting on southern hemisphere palaeoclimates held in Swaziland in 1983 (Vogel 1984).

In contrast to the pluvial theory, the astronomical theory demands that the basis for identification and correlation of Quaternary climatic changes is temperature rather than rainfall, and the key sequences are to be found in deep sea cores and not on land. The deep-sea core sequences have been correlated by Shackleton and Opdyke (1976), and others, and are divided into warmer phases (odd-numbers) and cooler phases (even-numbers). The Present Interglacial is Oxygen Isotope Stage 1, the Last Glacial Maximum is Stage 2, relatively warmer and cooler stages are Stage 3 and 4, and the Last Interglacial is Stage 5a–e (Fig. 1.3). This most recent interglacial/glacial/interglacial cycle is the best known but little work has been possible in Southern Africa to extend the palaeoclimatic chronology back much beyond this time range.

Attempts to link the Southern African land-based palaeoclimatic sequence to the late Pleistocene global temperature changes derived from oxygen isotope data in deep sea cores were made by Butzer (1978) in correlating sedimentological observations in the Klasies River Mouth caves on the southern Cape coast with sea level changes nearby. He interpreted the stratigraphy at the sites as indicating that Middle Stone Age occupation began there late in the Last Interglacial (oxygen isotope stage 5e) and continued through to the beginning of stage 4 when the sea level dropped well below its present level. Inland, Beaumont (Beaumont *et al.* 1978) made similar use of this chronology for the dating of the Middle and early Later Stone Age sequence at Border Cave. The temptation to apply the deep-sea core sequence to land-based data is great, but we need to encourage the build-up of an independent land-based chronology if we want to establish a comparable record of the effect of climatic forcing on terrestrial ecosystems.

The 1980s have seen experimentation with more sophisticated climatic models based on present-day circulation patterns that acknowledge a more complex interaction of atmospheric circulation than was previously envisaged (Flohn 1984; Harrison *et al.* 1984; Tyson 1986). In East and North Africa the models have focused again on precipitation, as evident from lake levels, as the primary data base (Nicholson and Flohn 1980; Harrison *et al.* 1984), but this method has not been successful for

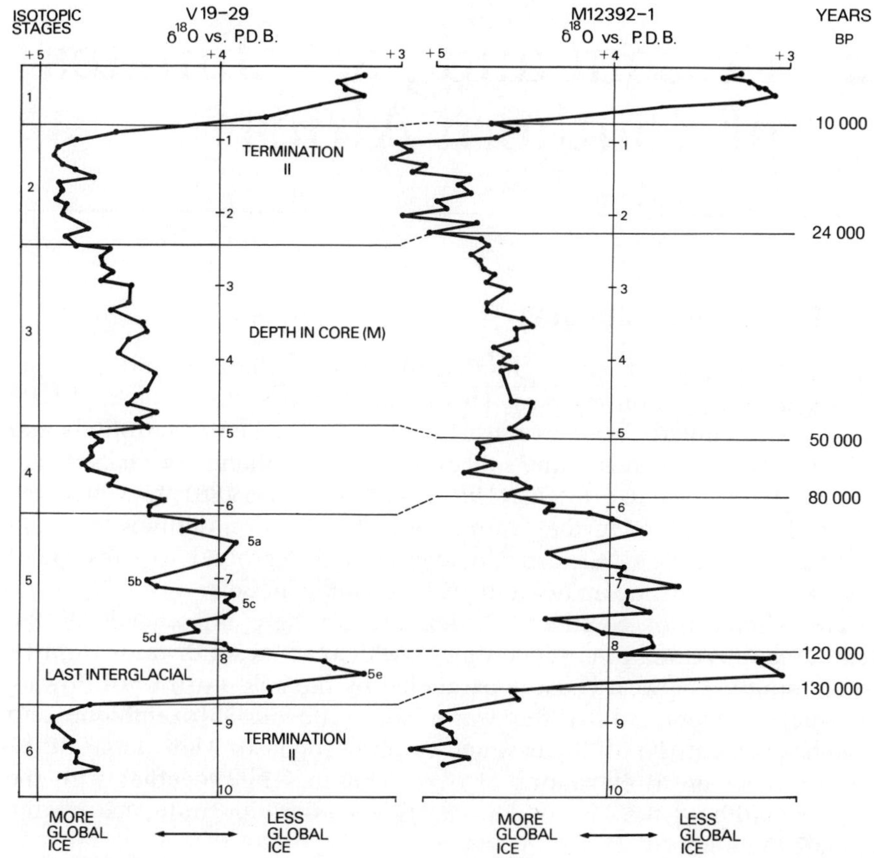

Fig. 1.3 Oxygen isotopic records of the last 150 000 years from Pacific core V19-29 and Atlantic core M12392-1, oxygen isotope stages and substages, and terminations (after CLIMAP Project Members 1984).

Southern Africa where there are fewer lakes with comparable information. These and other models are reviewed in greater detail in Chapter 2.2.

2 Present and past climates of Southern Africa

2.1 Present day climates

Today, Southern Africa is dominated by dry climates and strongly seasonal precipitation regimes. This is the datum against which all Late Quaternary climatic changes must be viewed. Detailed descriptions and analyses of the climates of the subcontinent are contained in Jackson and Tyson (1971), Schulze (1972, 1984), and Tyson (1986), to which the reader is referred for further information. This section outlines the main features of Southern African climates as a background to subsequent discussions of palaeoclimates and palaeoclimatic models.

The climates of Southern Africa are strongly influenced by the latitudinal position of the subcontinent, which tapers southwards from 18 to 35° south. The west coast is paralleled by the cold northward flowing Benguela current, whilst the warm south flowing Mozambique and Agulhas currents lie off the east and south-east coasts. Many areas of the subcontinent lie at elevations of over 1000 m. This, together with the narrow width of the land mass, exerts a moderating influence on the climate (Schulze 1972).

Southern Africa lies almost entirely within the subtropical high pressure belt which is separated by the land mass of Africa into two cells: the South Atlantic and Indian Ocean anticyclones. The mean circulation over Southern Africa is therefore anticyclonic throughout the year above the atmospheric boundary layer (Tyson 1986). North of 20° south, climates are strongly influenced by the seasonal movement of the Inter-tropical Convergence Zone (ITCZ). The circumpolar westerlies and their associated temperate low pressure disturbances lie to the south of the subcontinent. Southern Africa therefore lies in a zone of interaction between tropical and temperate circulations which influences climates at all seasons.

2.1.1 Circulation patterns

The South Atlantic and Indian Ocean subtropical anticyclonic cells are centred on 30° south, with 5–6° of seasonal latitudinal movement. The

Indian Ocean high tends to move south-eastwards in summer, whilst the South Atlantic cell weakens and moves north-west in winter. In summer a weak thermal low pressure cell may develop over the central interior of the continent and is linked by a trough to a tropical low north of Botswana (Taljaard 1981). The position of this low varies longitudinally along the southern branch of the ITCZ in the region of the Zaire Air Boundary and occasionally moves south to northern Botswana and Namibia. There is also a weak ridge of high pressure over the northern Cape Province. In winter, the mean circulation at the surface of the Southern African plateau is strongly anticyclonic, with a single large high pressure cell developed over the north-eastern Transvaal. The intensification and equatorwards extension of the westerlies reaches a maximum at this time of the year. The change to a single high pressure cell occurs in March, giving rise to an influx of moist tropical air over the western parts of Southern Africa and an autumnal rainfall maximum in these areas.

2.1.2 Climatic elements

Rainfall

Figure 2.1. shows the distribution of annual precipitation over Southern Africa. Rainfall is at a maximum along the eastern escarpment zone and decreases westwards. A second rainfall maximum occurs in the south-western Cape. The orographic effects of the Drakensberg escarpment and the Cape Fold Belt are very marked. In the south-western Cape, rainfall increases from 400 mm on the Cape Flats to over 2000 mm in the adjacent mountain ranges. The lowest rainfall totals are recorded along the Namib coast where less than 20 mm falls per year. The 500 mm isohyet divides Southern Africa into a wetter eastern and a dry western half.

Precipitation over the northern and eastern parts of Southern Africa occurs mostly in summer and more than 80 per cent of rainfall in these areas occurs between October and March. In the south-western Cape, over 80 per cent of rainfall occurs in the winter months of April to September. The contribution of winter rainfall to the annual total decreases rapidly north-eastwards. Only the narrow southern Cape coastal region and the adjacent mountain ranges receive all season rainfall, with equinoctial maxima.

Rainfall over the summer rainfall zone is mainly convectional in origin. Conditions favouring the development of convectional cells are controlled not only by diurnal heating and mesoscale forcing, but also by synoptic conditions which give rise to strong uplift of unstable moist air masses. A common synoptic scenario for the development of large scale convective activity is the occurrence of a trough in the tropical easterly

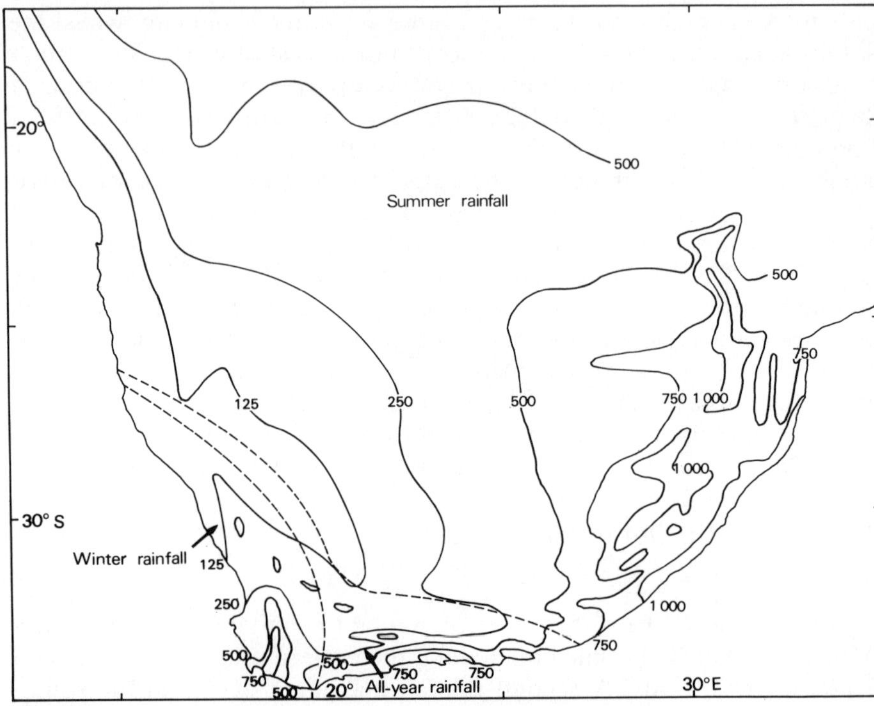

Fig. 2.1 Simplified mean annual rainfall (mm) for Southern Africa (after Schulze 1972).

circulation (Taljaard 1981). The passage of frontal systems also may trigger storms, but frequently no single synoptic scale forcing mechanism may be identified (Tyson 1986). In the winter rainfall zone, most rainfall is associated with the passage of cold fronts.

Periods of fine and dry weather over Southern Africa are associated with the development of large anticyclonic cells over the subcontinent with their attendant strong subsidence and temperature inversions. Such conditions are most common in winter, but they may also occur in summer. If they persist for extended periods at this season, heat waves and severe desiccation are the result (Jackson 1961).

Rainfall totals vary from year to year, with the interannual variability rising from 20 per cent over the wetter eastern areas of Southern Africa, to 60 per cent in the south-western Kalahari and more than 80 per cent in the central Namib. Tyson (1986) has identified a clear spatial pattern of extreme wet years (rainfall exceeding 125 per cent of normal), which are more common in the drier parts of the region and extreme dry years (rainfall less than 75 per cent of normal) which are more frequent in wetter areas. In the summer rainfall area, Tyson and his co-workers

(Tyson *et al.* 1975; Tyson and Dyer 1975, 1978; Tyson 1986) have recognized a weak 18-year periodicity in rainfall totals, comprising 9-year spells of predominantly wet years, followed by nine predominantly dry years. There is a strong spatial patterning to rainfall periodicities (Tyson 1986) with a 10–12-year oscillation being noted in the southern Cape. Over north-western parts of South Africa and Namibia, the oscillation is a 6-yearly one. The west central interior shows a quasi biennial cycle.

Surface temperatures

Compared to areas at similar latitudes (for example, North Africa and Australia), Southern Africa experiences relatively low temperatures as a result of the general elevation of the subcontinent and its proximity to oceanic influences. The range of temperatures (Fig. 2.2) is considerable and local topographic effects are important. The highest temperatures are recorded from the southern Kalahari, Mozambique, and western Zimbabwe where mean annual temperatures reach 20–24°C. The lowest

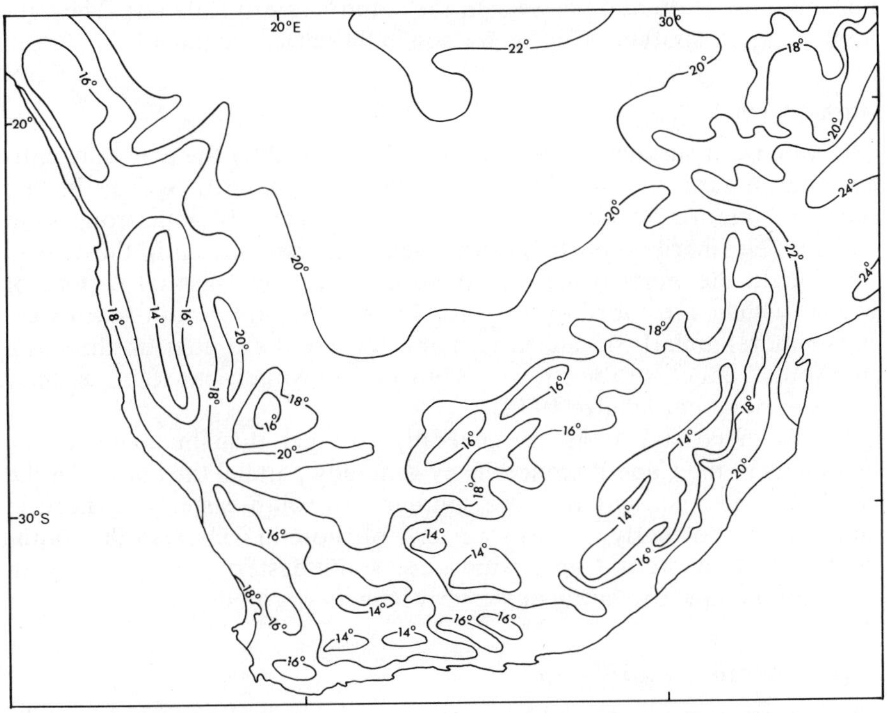

Fig. 2.2 Simplified mean annual temperature (°C) for Southern Africa (after Schulze 1972).

temperatures (11–12°C mean annual temperature) occur in the eastern highlands and Cape mountains. The importance of oceanic effect is shown by a comparison between the mean annual temperatures at 29° south, which are 15°C on the west coast and 21°C on the east coast. The greatest diurnal temperature ranges are found in the interior, especially in the Kalahari and Karoo. The highest frequency of temperatures below freezing occurs along the Natal–Lesotho mountains and escarpment zones where there may be more than 90 days a year when the temperature falls below 0°C.

Humidity and evaporation

The major moisture source for most of Southern Africa is the Indian Ocean and there is a strong east to west gradient in atmospheric moisture content, especially in summer (Jackson 1961). Relative humidity at 14.00 hours declines from 70–80 per cent in coastal areas to 50 per cent in eastern plateau areas, and less than 30 per cent in the west (Schulze 1972).

Pan evaporation rates increase inland and from east to west, reaching a maximum of 4000 mm per year in the south-western Kalahari. They are least along the southern Cape coast and in the eastern highlands.

Surface winds

The pattern of surface winds is strongly influenced by the generally anticyclonic nature of circulation patterns in Southern Africa. In the northern interior, winds are easterly to north-easterly and strongest in summer. Northerly to north-westerly winds prevail all year in the central interior. In the western interior, there is a strongly seasonal pattern of winds which are north-north-east in winter, but south-south-west in summer. Similarly, in the eastern and south-eastern interior, there is a monsoonal effect with north-westerly winds in winter, but south-easterly winds in summer (Schulze 1972).

Winds in coastal areas are generally stronger than in inland areas, especially in the Cape Province. They generally parallel the coast. On the south and east coasts winds are southerly to south-easterly onshore in summer and northerly or north-westerly offshore in winter. In the south-western and western Cape, winds are south-easterly or southerly in summer, but north-westerly or westerly in winter.

Causes of rainfall variations

The synoptic conditions which give rise to extended wet or dry spells have been discussed by Harrison (1986) and Tyson (1984, 1986). In the summer rainfall zone, wetter spells on the scale of days, seasons and years are associated with the development of lowered pressure over Southern Africa and increases in pressure over the South Atlantic. In dry spells, the converse applies. The development of such circulation patterns appears to be linked to variations in the zonal Walker circulation, the so-called Southern Oscillation. In the high phase of the Oscillation, tropical easterly flow strengthens and the westerlies contract polewards, but become stronger around latitude 47° south. During the dry (low) phase, the westerlies weaken polewards, but expand and strengthen northwards and the tropical easterlies weaken (Tyson 1986). Meridional circulations over Southern Africa and the adjacent oceans, together with interactions between tropical and temperate perturbations, provide the link between the African Walker cell and the subtropical circulations. The major form of interaction between tropical easterly and temperate westerly circulations occurs in the summer months and takes the form of extended cloud bands, visible on satellite images, which link tropical easterly waves and temperate westerly waves and low pressure systems. Enhanced convective activity and rainfall over the interior of Southern Africa is usually associated with the cloud bands.

On the basis of these observations, Tyson and his co-workers (Tyson 1986) have put forward a model for the atmospheric circulation in extended wet and dry spells. Most rainfall over the interior of the subcontinent results from the favourable conjunction of deep tropical and westerly disturbances, which allow the formation of major north-west–south-east orientated cloud bands. During wet spells (Fig. 2.3), the ascending part of the African Walker cell is located over tropical Africa. The subtropical jet stream is strengthened to the east of the subcontinent. The ITCZ is also strengthened, resulting in more easterly waves and enhanced easterly flow and advection of moisture from the Indian Ocean. Changes in the amplitude of the upper level Atlantic wave result in the preferred location of the cloud bands over Southern Africa. At the same time, there is a southward shift of storm tracks because of higher poleward energy fluxes and a lower meridional temperature gradient, resulting in drier winters in the winter rainfall zone.

In dry spells (Fig. 2.3) the rising limb of the African Walker circulation is situated off the coast of Africa, the ITCZ is weakened and tropical easterly waves diminish. Pressure rises over Southern Africa and moisture advection to the subcontinent decreases. The South Atlantic anticyclone is weaker and the Atlantic wave diminishes in amplitude, such that cloud bands locate preferentially over the Indian Ocean. Poleward energy fluxes decrease, but temperature gradients increase, leading to an equatorwards movement of storm tracks, bringing more winter rain to the south-western Cape.

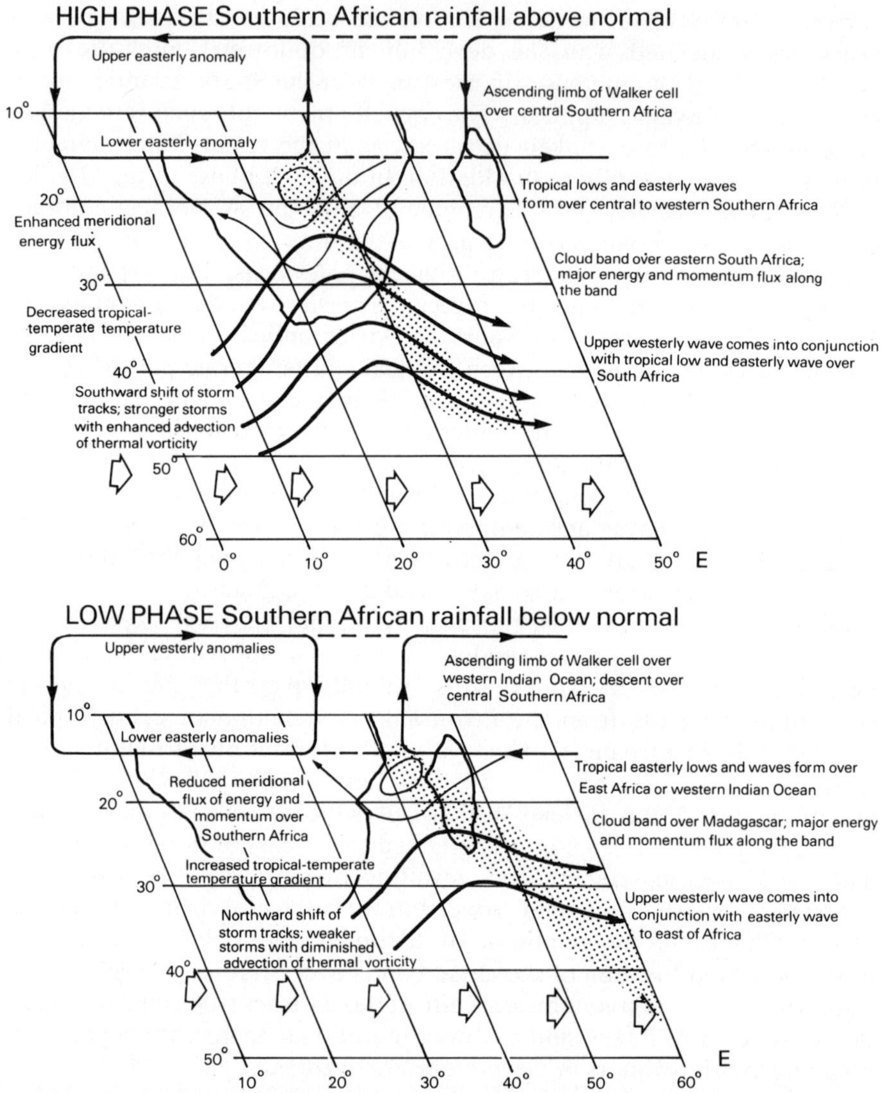

Fig. 2.3 Schematic representation of the anomalous Walker circulaton over Southern Africa during high phase of the Southern Oscillation (above) resulting in above-normal summer rainfall; and during low phase (below) resulting in below normal summer rainfall and above normal winter rainfall (after Tyson 1986, fig. 9.4).

2.2 Palaeoclimatic models

Global circulation models have been calculated for the Last Glacial Maximum and in this section the results will be compared with atmospheric circulation models that have been proposed for Southern Africa. The latter can be usefully divided into models that assume:

(a) an equatorward movement of circulation belts in both hemispheres;
(b) stronger circulation patterns without latitudinal displacement; and
(c) a longitudinal displacement of circulation patterns.

All three tend to focus on a few features of atmospheric circulation rather than taking into account all the possible variables, but this has been a necessary device for testing expectations against the field data. In this section we focus on the explanations proposed for Last Glacial Maximum conditions.

2.2.1 Global circulation models for the Last Glacial Maximum

By the early 1970s sufficient data had been accumulated from deep sea cores to allow some correlation between these and modern climatic data gathered around the world. In 1971, the influential CLIMAP (Climate, Long-range Investigation, Mapping, and Prediction) programme was launched to reconstruct from quantitative geological evidence the average state of climatic boundary conditions (sea-surface temperature, ice extent, ice elevation, continental albedo, wind velocity, rainfall/precipitation, land surface temperature, etc.) for the Last Glacial Maximum at 18 000 BP, and to measure the oscillations of Pleistocene climate worldwide. As part of this programme, general circulation models (GCMs) were developed which mathematically simulate ice age climatic conditions by explaining in dynamic terms the complex of processes which maintain a particular climate in equilibrium (CLIMAP 1976, p 1131).

GCMs are computed from equations governing the transfer of heat and momentum in the atmosphere based on specifications for individual locations of incoming solar radiation, land topography, sea surface temperature, surface albedo, and ice sheet extents and elevations. These models can then be compared against the evidence gathered from biological and geological data. The general acceptability of the results of such a simulation are shown by Heath (1979, fig. 1) who contrasts a simulated model for present-day conditions and a map summarizing observed present-day precipitation around the world. The correspondence is reasonable, but not absolute. Simulated models of boundary conditions for the Last Glacial Maximum have been computed for global sea-surface temperature (CLIMAP 1976), land surface temperature, surface heat balance, sea level pressure (Gates 1976; Manabe and Hahn 1977), precipitation (Heath 1979), relative humidity, cloudiness, evaporation rate, evaporation patterns, and wind velocities (Gates 1976;

thein 1978). In a comparison of five simulation models, Williams 978) found considerable variation in temperature estimates for particular localities, but a general agreement that the Last Glacial Maximum was cooler and distinctly drier worldwide. These generalizations provide a broad standard against which field observations can be compared. Some of the GCMs for Africa are summarized in Fig. 2.4. Together, the CLIMAP data and GCMs confirm that 18 000 years ago temperatures were markedly cooler and that precipitation was generally lower. What is still at issue is just how much cooler and how much drier (if at all) it was at individual stations in Southern Africa. In order to assess this problem we review models proposed in the past to explain how and why Last Glacial Maximum climates differed from those of the present.

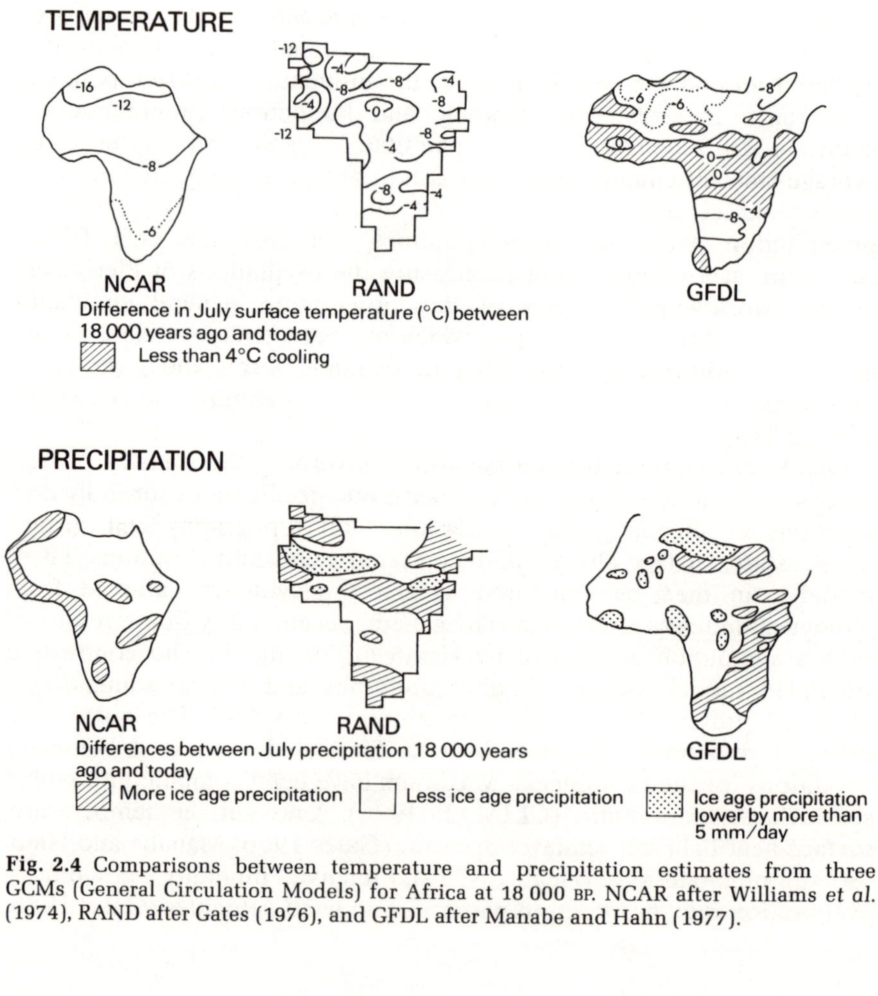

Fig. 2.4 Comparisons between temperature and precipitation estimates from three GCMs (General Circulation Models) for Africa at 18 000 BP. NCAR after Williams *et al.* (1974), RAND after Gates (1976), and GFDL after Manabe and Hahn (1977).

2.2.2 Models for Last Glacial Maximum climates in Southern Africa

It is inequalities in temperature which provide the energy for air circulation and, therefore, climate. The presence or absence of a polar ice cap on and around Antarctica has a significant effect on the southern hemisphere air and ocean temperatures, as well as on the wind and ocean currents, as can be shown in the scale of global climatic changes that took place before and after Antarctica moved into position over the South Pole (Van Zinderen Bakker 1976; H. J. Deacon 1983; Flohn 1984).

The Antarctic ice sheet at present contains 90 per cent of the world's ice and covers an area of 14 million km^2. In addition, the Antarctic sea-ice varies from 4 million km^2 in March, to 20 million km^2 in September. By contrast, the Arctic sea-ice cover has an extent of only 7–12 million km^2 in September and March/April, respectively. Thus, while the heat budget of the Arctic leads to minimum temperatures of −35°C, those in the Antarctic drop as low as −70° or lower (Pittock 1978; p 4), leading to a pole-to-equator temperature gradient which is 40 per cent larger over the southern hemisphere than over the northern hemisphere (Lamb 1972, p. 94). This anomaly results in mid-latitude westerly winds in the southern hemisphere being much stronger than those in the northern hemisphere. However, during a glacial maximum, this anomaly would be partly adjusted because of the much larger extension of ice over the northern continental landmasses, with a consequent increase in the northern hemisphere pole-to-equator temperature gradient. The strength of the present-day southern hemisphere westerlies leads to a displacement of the ITCZ 7° of latitude northward of the geographical equator in Africa (Flohn 1967, 1984; Pittock 1978).

The positioning and strength of the southern hemisphere westerlies are important aspects in the assessment of glacial maximum conditions for Southern Africa because they bring much of the winter rainfall to the western and southern Cape at present, and because strong westerlies help to maintain the zonality of the present-day circulation pattern by preventing the poleward migration of subtropical anticyclones (Boucher 1975, p. 280). The zone between the pack ice and the equatorward boundary of the westerlies, i.e. between the circumpolar trough and the subtropical ridge is dominated by cyclonic circulation, while anticyclones prevail to the north of the subtropical ridge. These two circulation patterns at present maintain the summer rainfall maximum in the interior, eastern and southern margins of Southern Africa on the one hand and the winter rainfall maximum on the western and southern margins on the other.

Initial interest in reconstructing past climatic changes in Southern

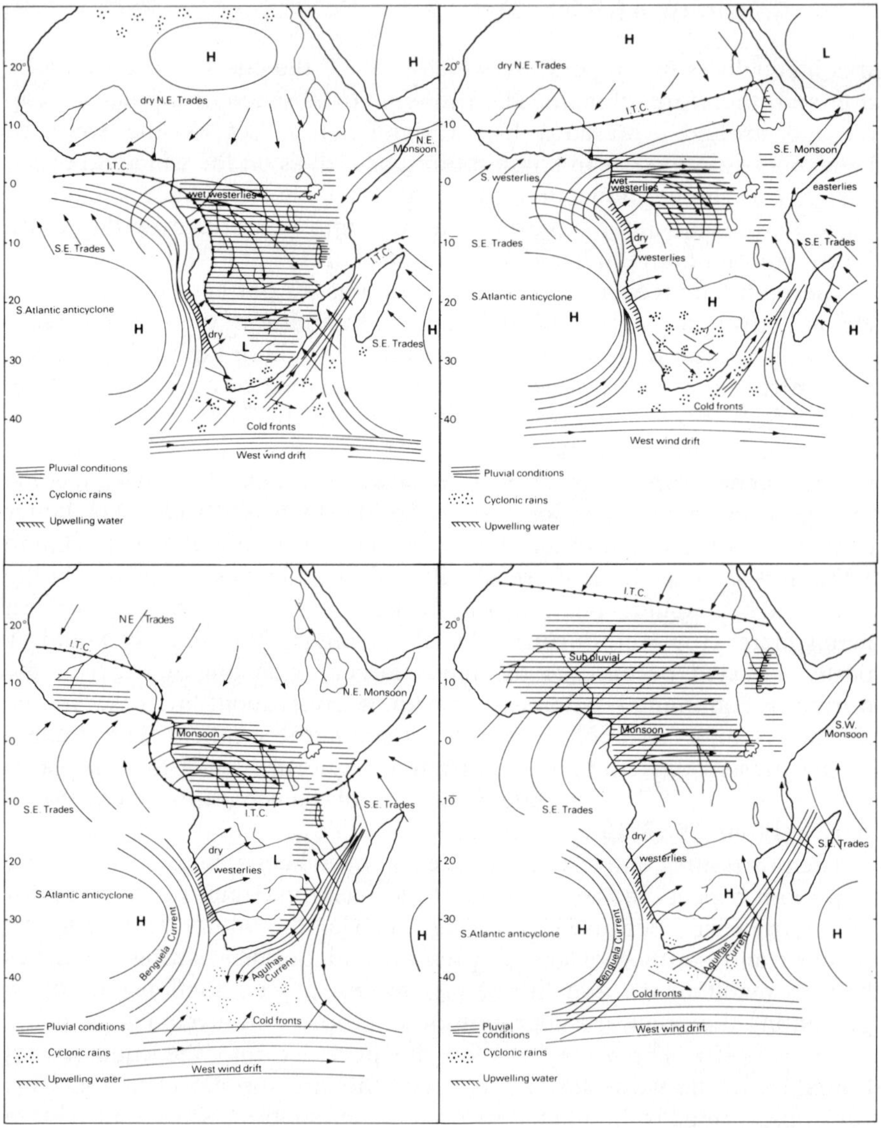

Fig. 2.5 Model of some palaeoclimatic trends in Africa proposed by Van Zinderen Bakker (1967). Top left: glacial southern summer; Top right: glacial southern winter; Bottom left: interglacial southern summer; Bottom right: interglacial southern winter. (Reproduced with permission from the author and The University of Chicago Press. © 1967 Wenner–Gren Foundation.)

Africa was stimulated mostly by the palynological work of Van Zinderen Bakker (1957, 1962*a*). His first comprehensive reconstruction of glacial/interglacial climates considered the African continent as a whole (Van Zinderen Bakker 1962*b*, 1967) and maps of the major circulation patterns were drawn up (Fig. 2.5). He based these on the assumption that during the Last Glacial Maximum the zonal climatic control on the latitudinal positions of the major circulation patterns was in general the same as at present, but because the polar high pressure cells extended further towards the equator, the climatic belts were assumed to have contracted, accelerating the wind and ocean circulation systems. This resulted in a southward displacement of the ITCZ. One of the major implications was that the winter rainfall region of the western and south-western Cape would have been shifted further north than at present (Van Zinderen Bakker 1976, p. 144), with the corollary that the vegetation typical of the Cape Ecozone would have extended northwards into South West Africa, Namibia, and into the Cape and Orange Free State north of the Cape Fold Mountains.

Reiterating this model some years later, Van Zinderen Bakker (1976) (Fig. 2.6) cited confirmation for the hypothesized northward shift of the climatic belts in evidence for grassland habitats in the southern Cape, for archaeological proxy evidence which indicated that many sites in the southern Cape had no occupation deposits dating to the Last Glacial Maximum, and for geomorphological evidence which indicated much cooler temperatures in the Drakensberg. Data were also available at that

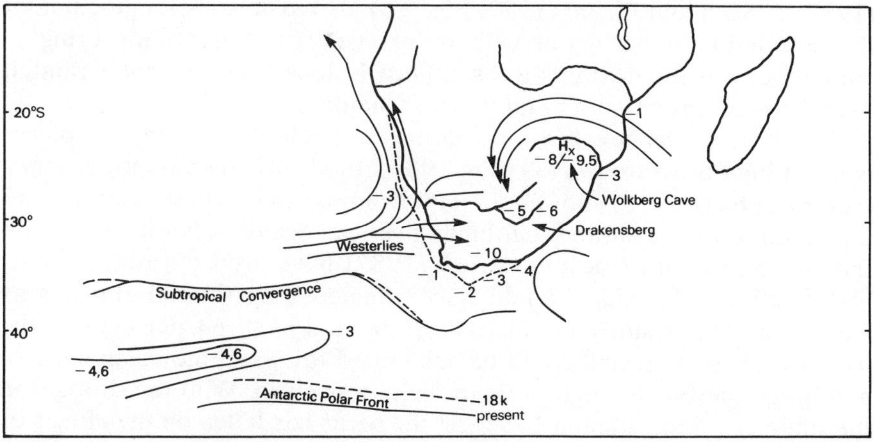

Fig. 2.6 Model of palaeoclimatic trends in Southern Africa at *c.* 18 000 BP from Van Zinderen Bakker (1982*a*). Ocean temperature anomalies after Newell *et al.* (1981). Position of subtropical convergence and Antarctic Polar Front after Morley and Hays (1979). Kalahari wind directions for summer and winter. (Reprinted from: *Palaeoecology of Africa*, (ed. J. C. Vogel, E. A. Voigt, and T. C. Partridge) Volume 15. Southern African Society for Quaternary Research, 1982. A. A. Balkema, P.O. Box 1675, Rotterdam, Netherlands.)

time to show that ocean temperatures were between 2 and 5°C cooler than at present off the Southern African coasts, and the palaeotemperature record of a stalagmite from the central Transvaal indicated a glacial maximum temperature of 8–9.5°C lower than at present (Talma *et al.* 1974). Van Zinderen Bakker (1976, p. 169) proposed, then, that glacial maxima were marked by dry and cool conditions in the north and east of Southern Africa, and cold temperatures, strong winds, and wet winters in the southern section. This hypothesis was of considerable value in giving a scenario for glacial and interglacial climates that could be tested in the field (H. J. Deacon 1983).

In a more detailed model that focused on Africa north of the equator, Nicholson and Flohn (1980) suggested an equatorward displacement of circulation features in the southern hemisphere that would have led to higher rainfall during the Last Glacial Maximum in the region between about 25 and 30° south, drier conditions to the north and about the same rainfall as today in the extreme south (Fig. 2.7).

Another model, proposed by Butzer *et al.* (1978*b*) stresses stronger circulation. It suggests that at glacial maxima, with the westerlies positioned at much the same latitudes as at present and an ITCZ southward of its present position, warm moist tropical air from the Congo and the Indian Ocean would have been advected more often over the summer rainfall region bringing higher rainfall, whereas the winter rainfall region would have received less rain except perhaps for the extreme south-west. A variation proposed by Van Zinderen Bakker (1980, 1982*a*) and Heine (1982; Fig. 2.8) also assumes strengthening of the circulation pattern, but with more vigorous trade winds bringing winter rains as far north as the southern Kalahari, and summer rainfall over the Kalahari, but not as far west as Namibia.

The third model invokes longitudinal instead of latitudinal displacement of high pressure cells (Tyson 1986). In examining modern weather patterns, Dyer (1976) showed a relationship between annual rainfall and the position of the subtropical highs over the South Atlantic and South Indian oceans, and Cockcroft *et al.* (1987) have used the observations to hypothesize how this would have affected Last Glacial climates as well. In a similar study in Australia (Pittock 1973), 84 per cent of the total variance in rainfall could be accounted for when the mean latitude and longitude for the high pressure belts was given. With the drought in the summer rainfall area in 1982–83 the focus has fallen on the effects of the Southern Oscillation or the related 'El Nino' effect in which the Southern Indian ocean high pressure cell is displaced northeastwards of its 'normal' position bringing with it much drier conditions in the summer rainfall region of Southern Africa. The implication is that such conditions may have prevailed over extended periods. Pittock and Salinger (1982) have proposed that just such an extended Southern

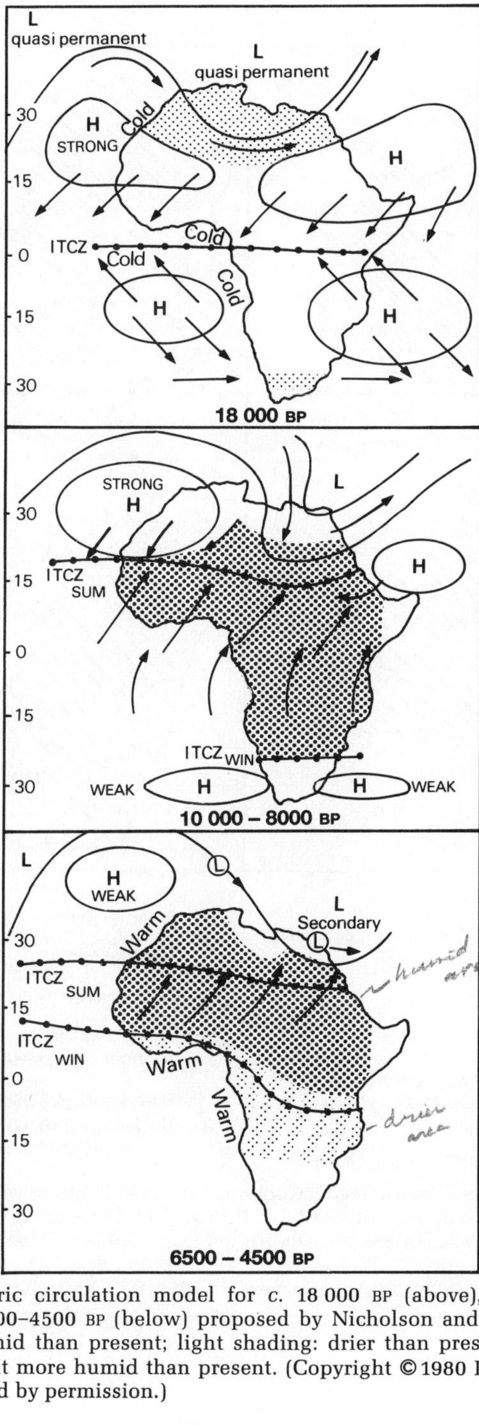

Fig. 2.7 Atmospheric circulation model for *c.* 18 000 BP (above), *c.* 10 000–8000 BP (middle) and *c.* 6500–4500 BP (below) proposed by Nicholson and Flohn (1980). Dark shading: more humid than present; light shading: drier than present; hatching: drier than previously, but more humid than present. (Copyright © 1980 D. Reidel Publishing Company. Reprinted by permission.)

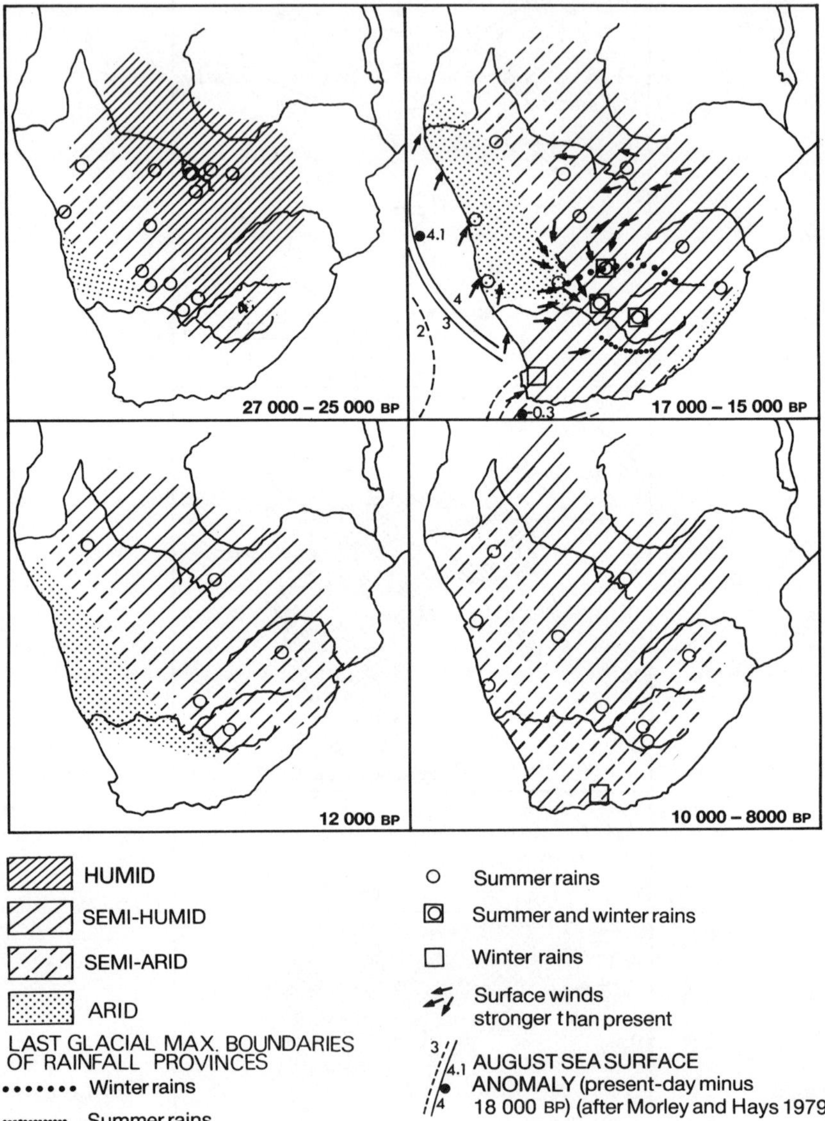

Fig. 2.8 Hypothetical climatic reconstructions for the Kalahari at various times between 27 000 and 8000 years BP, proposed by Heine (1982). Based on radiocarbon dated deposits, molluscs, cave sinters, etc. (Reprinted from: *Palaeoecology of Africa*, (ed. J. C. Vogel, E. A. Voigt, and T. C. Partridge) Volume 15. Southern African Society for Quaternary Research, 1982. A. A. Balkema, P.O. Box 1675, Rotterdam, Netherlands.)

Oscillation could account for the climatic changes observed during the Last Glacial Maximum in Australia and New Zealand, and Salinger (1984), Cockcroft *et al.* (1987), and Tyson (1986) have suggested that a Southern Oscillation pattern could explain Southern African Last Glacial Maximum climates too.

On the basis of present-day synoptic patterns and rainfall variations, they suggest that the summer rainfall region in Southern Africa was drier and the winter rainfall region was wetter during the Last Glacial Maximum. At this time the Southern Oscillation (Low Phase) was enhanced over a period of several thousand years, coinciding with an eastward movement of the Walker circulation over the South Indian Ocean and Malagasy, and along the west coast, a northward movement of the boundary between the winter and summer rainfall regimes from its present position at 30° south to about 25° south (Fig. 2.3). Cockcroft *et al.* (1987) base this latter estimate on the results of Howard and Prell (1984), and Howard (1985) which suggest that the Subtropical Convergence between Africa and Australia shifted between 3 and 5° equatorwards of its present position. This is at variance with the findings of Morley and Hays (1979) who saw a much smaller shift of the Subtropical Convergence in the South Atlantic Ocean during the Last Glacial Maximum. This anomaly may make it necessary to re-evaluate the suggestion that the winter rainfall boundary moved significantly northwards, particularly in view of the absence of good evidence for higher rainfall in that region during the Last Glacial Maximum.

Other factors that have not played a major part in the models discussed above but would have affected weather and climatic patterns of all kinds during glacial times should also be taken into account. For example, Tyson (1977) has suggested that cool periods are generally correlated with enhanced climatic instability and an increase in meridional blocking leading to greater extremes of climate. In Australia, summer temperatures in the Last Glacial Maximum are estimated to have been hotter and winters cooler than the present. Combined with greater wind speeds, this led to drier and generally more extreme weather conditions over much of the continent (Bowler 1976; Ash and Wasson 1983). In addition, several researchers (Gates 1976, p. 1143; Williams 1978, p. 192; Sarnthein 1978; Heath 1979) have pointed out that cooler oceans would have led to lower evaporation rates with a resultant drop in the amount of moisture available for precipitation. Gates (1976) has estimated that this would have been about 23 per cent lower in July in the southern hemisphere and 37 per cent lower in the northern hemisphere 18 000 years ago. A reduction in the amount of moisture evaporated from cooler oceans could have been felt in the winter rainfall region where the winter cyclones could have been both windier and drier. In the all-year rainfall belt on the southern Cape coast a weaker Agulhas current (Martin 1981) could have led to cooler ocean temperatures and the exposed continental shelf would have placed the coastline 80–100 km south of its present position thereby reducing the effect of orographic rainfall that at present accounts for a significant proportion of the annual precipitation along the southern Cape coast.

2.2.3 Test implications for palaeoclimatic models

There is some variability in the reasons for the reconstructions made by the three models, but all agree that Last Glacial climates were probably wetter, windier, and cooler in the winter rainfall region which extended further north and east than at present, and that the summer rainfall region was cooler and probably drier. Where mid-Holocene climates are discussed, the situation is reversed.

Most researchers are in agreement that the concept of an equatorward displacement of circulation belts is too simplistic an explanation, so the question is whether the stronger circulation model or the model that proposes a long-term Southern Oscillation and longitudinal displacement of the Walker circulation is the more appropriate. The implication is that if the former model is correct, then the summer rainfall region would have been moister than if Southern Oscillation drought conditions had pertained. In both models, it is necessary to demonstrate that higher winter rainfall was received north and east of the present-day winter rainfall region, and that the summer and winter rainfall regions were out of phase.

3 Sources of palaeoclimatic evidence in Southern Africa

3.1 Introduction

Ongoing research is expanding the data base for palaeoclimatic studies and there is a broader range of information than there was a decade ago. This is noticeable particularly in the biological sciences. Isotope studies of speleothems, aquifers, and fossil bones have begun to yield particular details unavailable from other sources, and new dating techniques are being employed where previously only radiocarbon was available. Studies of landforms have been extended both geographically into regions previously unstudied, and with greater detail using Landsat imagery and X-ray diffraction for sediment analyses. The types of evidence and the advantages and problems of interpretation are discussed in detail below. For the most part they remain, as Flint pointed out in 1959, second order derivations with climate inferred from the landforms, sediments, or plant and animal remains.

3.2 Geomorphic and sedimentary sources

Much of the evidence that has been used to reconstruct Quaternary palaeoclimates in Southern Africa has been obtained from studies of landforms and sediments. This is a consequence of the physiography of the sub-continent, with its large areas of low relief and its widespread arid to semi-arid climates. There are few large sedimentary basins and little opportunity for the accumulation of organic rich sediments. Often sedimentary sequences are short and geographically scattered making correlations unsatisfactory, and landforms frequently provide the best means of establishing regional patterns of climatic change.

As a prelude to subsequent discussions of the evidence for palaeoclimates on a regional basis, it is necessary to examine the characteristics of the main types of geomorphic and sedimentary evidence, and their limitations for providing palaeoclimatic information. It should be emphasized that frequently palaeoclimatic reconstructions are made on the basis of evidence from several sources, and that many of the types of evidence discussed below are complementary to each other.

3.2.1 Aeolian sediments and landforms

Mobile or active continental sand dunes exist today, where sand supply conditions permit, in areas where mean annual rainfall is 150–100 mm or less. The existence of such dunes is thus an unequivocal indicator of aridity and the existence of fixed or vegetated dunes in areas where rainfall is above this limit provides evidence for the former distribution of arid climates. Such systems of dunes have been described from Southern Africa by Lancaster (1981), and Thomas and Goudie (1984), and from Australia (Jennings 1968), the southern Sahara (Grove and Warren 1968; Talbot 1980), India (Goudie *et al.* 1973), and central Asia (Petrov 1975).

Comparisons of the limits of presently active and fixed dunes may indicate the magnitude of shifts in climatic belts (e.g. Warren 1970). It is now clear that such comparisons are over-simplified and the effects of interactions between wind strength, vegetation, and sand mobility are important, as discussed for the Simpson Desert in Australia by Ash and Wasson (1983) and for the Sahel by Talbot (1984).

At a regional scale, the pattern of dunes reflects the regional pattern of winds capable of moving sand. Comparison of modern winds and potential sandflows with those indicated by fixed dunes may provide evidence for changes in wind regimes and circulation patterns (e.g. Warren 1970; Sprigg 1979; Lancaster 1981).

In some areas dunes of different types and alignments may be superimposed, affording the potential for distinguishing different periods of dune formation (Sprigg 1979; Fryberger 1980; Mallick *et al.* 1981).

Lake margin dunes, such as those adjacent to many pans in the Kalahari, form in particular hydrologic and climatic conditions and can be used to provide valuable data on local and regional lacustrine environments (Bowler 1971; Lancaster 1978*b*).

Less certain is the palaeoclimatic significance of river source dunes such as those described by Van Rooyen and Burger (1974) from the north-western Orange Free State and coastal dunes, which are widely distributed along the Natal and Cape coasts (Illenberger 1986).

3.2.2 Alluvial sediments and landforms

Former river channel and floodplain deposits are often preserved as river terraces or buried channelways adjacent to modern river valleys.

Sequences of such deposits and landforms may provide evidence for the geomorphic evolution of the valley and indirectly contribute palaeoenvironmental information. The sedimentary character of the deposits may indicate the flow conditions under which they were laid down. River terraces and their deposits represent periods of floodplain aggradation. They are often separated by phases of erosion and incision. Either incision or aggradation may occur in both humid or arid climates and the behaviour of a river is not necessarily determined by changes in regional climates. Rather, the dynamics of floodplains are controlled by an interplay of the supply of sediment, its size and sorting character, and the magnitude of the channel forming discharges (Schumm 1968). These factors are linked to climatic changes in a way that is far from understood, particularly in Southern Africa (Butzer *et al.* 1973*b*). Helgren (1979) documents some of the possible reactions of stream channels to climatic change, emphasizing not only the nature of the change, but its direction.

Thus, despite the many studies of river terraces and their deposits which have been made in Southern Africa, their contribution to palaeoenvironmental knowledge has been limited and the interpretation of the evidence controversial. The classic area of the lower Vaal River has received much attention following the pioneer work of Van Riet Lowe (Sohnge *et al.* 1937) and continuing with the reinterpretations of Partridge and Brink (1967), and the detailed investigations of Butzer *et al.* (1973*b*) and Helgren (1979).

Bond (1967) summarized investigations of fluvial deposits in Zimbabwe, questioning the palaeoclimatic significance of many of them. In the Namib, studies of fluvial deposits by Rust and Wienecke (1974, 1980), Marker (1977*a,b*), Ward (1982, 1984*a,b*) and Lancaster (1984*a*) have assisted in the elucidation of the geomorphic history of the area, but provided little precise palaeoclimatic information.

Studies of drainage networks, now largely dry, in arid and semi-arid areas, may provide evidence for periods of greater runoff. The extensive fossil valleys of the Kalahari (Omaramba in Namibia, Mogacha in Botswana) imply considerably greater run-off in former times. In the Namib, the contraction and blocking by sand dunes of such rivers as the Tsondab and Tsauchab may be evidence for the increasing aridity of that region.

3.2.3 Palaeolakes

The identification of former high level shorelines and the study of lacustrine deposits provides evidence for hydrological changes in lake basins. As changes in catchment hydrology can be linked to climatic changes, studies of palaeolakes often provide valuable palaeoclimatic data. Particularly with closed lake basins, studies of past and present water balances can be made and estimates of former precipitation and evaporation provided. However, as Street-Perrott and Harrison (1985) have remarked, a more cautious and critical mood has replaced the initial phase of euphoria about the value of lake level fluctuations as a clear indicator of past climatic changes. This has resulted from the discovery of various complications, most notably those associated with ground-water and anthropogenic effects.

The existence of seasonal or permanent water bodies in presently arid areas, and evidence for high or low levels in present lake basins nevertheless continues to be one of the most useful and reliable indicators of Quaternary climatic changes in low latitudes. Correlations of absolutely dated shorelines and deposits have done much to increase our knowledge of the extent and duration of late Pleistocene humid periods (Street and Grove 1979). Lake basins are important sediment traps, and studies of cores through lacustrine sediments frequently provide important palaeoclimatic information (e.g. Coetzee 1967; Gasse and Street 1978).

Unfortunately, there are few large palaeolake basins in Southern Africa. A major palaeolake existed in the Makgadikgadi depression (Grove 1969; Grey and Cooke 1977). A further example probably occurred in the Etosha basin in Namibia (Rust 1984). Throughout the Kalahari, there were small palaeolakes in the pan depressions (Lancaster 1978*a*). In the Namib Sand Sea small pans occur in interdune areas (Teller and Lancaster 1986*a*). Their small size has facilitated detailed studies of sediments which have yielded useful palaeohydrological information. Pans are widespread throughout the continental interior and when they have been studied in detail can provide important data (e.g. Butzer *et al.* 1973*a*; Butzer 1984*b*; Partridge and Dalbey 1986).

3.2.4 Caves, cave deposits, and karst landforms

The protected environments of caves, whether coastal or inland, frequently give rise to sites of significant sediment accumulation. Analyses of their sediments, and associated faunal and archaeological material, often by multidisciplinary teams, have provided palaeoclimatic information for many areas of Southern Africa. The frequent associations of archaeological and faunal remains with cave sediments may provide material for absolute or relative dating, and facilitate correlations with other sites.

Analyses of cave sediments in the southern and south-western Cape (e.g. Butzer 1973*a*, 1978, 1979; Tankard and Schweitzer 1974; Tankard

1976*a*; H. J. Deacon 1979) have provided much of the information on palaeoclimates in this region. Elsewhere, the studies of Cooke (1975) at Drotsky's Cave and Butzer *et al.* (1978*a,b*) at Border Cave have yielded important palaeoclimatic information for areas which otherwise lack data.

Karst landforms are common on outcrops of limestones and dolomites throughout Southern Africa. Major areas which have been studied are the north-east Transvaal and the Gaap Escarpment. Marker (1972) discusses the evidence for Quaternary climatic change derived from these studies and examines some of the problems of its interpretation. Major problems relate to the supposedly diagnostic nature of certain landform assemblages and their restriction to specific climatic zones.

Caves in limestone areas may contain features and deposits of palaeoclimatic significance. Periods of cave formation under phreatic or vadose conditions may indicate changes in regional water tables which may be climatically determined. The sediments of cave fills can yield information on environments within and outside the cave (e.g. Brain 1958, 1967; Cooke 1975; Partridge 1978). Speleothems may also provide material for isotopic dating and determinations of palaeotemperatures (Talma *et al.* 1974; Vogel 1983).

The carbonates dissolved in the groundwaters of karst terrains as a result of solution of limestone are frequently redeposited at the sites of groundwater discharge. Large tufa sequences occur in the north-eastern Transvaal and along the Gaap escarpment. In this locality they form a depositional sequence extending throughout the late Cainozoic, and provide one of the most complete sedimentary and palaeoenvironmental sequences in Southern Africa (Butzer *et al.* 1978*b*).

The link between speleothem and travertine formation and climate has been tested worldwide by Hennig *et al.* (1983) who combined age data for 600 speleothems and 140 spring deposited travertines to show that their formation is stimulated by climate. Growth in general was low in the Last Glacial and highest in the Last Interglacial.

3.2.5 Periglacial landforms and deposits

Landforms and deposits formed under the influence of frost action are common today in high altitude areas in temperate and sub-tropical latitudes, including the mountains of Lesotho and the Drakensberg (Alexandre 1962; Hastenrath and Wilkinson 1972). The identification of such features at lower altitudes and latitudes may provide evidence for periods of lower temperatures during the Pleistocene glacial periods. However, their identification, interpretation and chronology is controversial, as discussed by Butzer (1973*b*). The significance of some deposits and landforms thought to be of 'periglacial' origin by earlier workers (Harper 1969; Linton 1969) have been reassessed (Butzer 1973*b*).

Frost phenomena and associated deposits may occur in cave fills, where they have been interpreted as good evidence for periods of accelerated frost weathering and colder climates (Butzer 1973*b*; Tankard 1976*a*).

3.2.6 Relict shorelines and associated deposits

The climatic fluctuations of the Pleistocene glaciations were accompanied by major changes in global sea levels as ice sheets grew and waned. Recognition of relict shorelines and their deposits may provide evidence on coastal palaeoenvironments, such as surface water temperatures (Tankard 1975). In addition, after due allowance has been made for the effects of regional tectonic movements, the altitude of the shorelines may afford the possibility of making correlations between deposits and landforms, either regionally or on a global basis. Changes in sea levels also affected the base levels of all coastal rivers, and need to be taken into consideration in assessing the palaeoenvironmental significance of fluvial deposits in their valleys (Dingle and Rogers 1972*a,b*).

Extensive systems of coastal dunes associated with high level shorelines occur in the southern and south-western Cape (Tankard and Rogers 1978; Illenberger 1986) and in Natal (Maud 1968; Davies 1976). They may provide evidence on the nature of coastal wind regimes and sediment sources.

The exact relationships between glacio-eustacy, global isostasy, and sea level changes have not yet been resolved satisfactorily (Aharon 1983; Butzer 1983; CLIMAP 1984; Kaufman 1986). For this reason, sea level changes along the Southern African coast need to be assessed independently of the scale and timing of changes in deep sea cores and shorelines elsewhere in the world so that correlations and discrepancies can be properly evaluated (see Cronin *et al.* 1981).

3.2.7 Duricrusts and palaeosols

Indurated, near surface or surface concentrations of iron oxides, calcium carbonate, silica, or gypsum forming, respectively, ferricrete, calcrete, silcrete, or gypcrete are a common feature of many sub-tropical and tropical landscapes. Their physical and chemical composition, and thus their distribution are strongly controlled by climatic factors acting through pedogenic processes (Goudie 1973). Thus, comparison of the distribution of duricrusts with present climatic patterns may provide some information on palaeoclimates (Mason *et al.* 1959; Birkeland 1974; Netterberg 1978; Yaalon and Ward 1982). Such an approach should be treated with caution, as the processes by which duricrusts form are still imperfectly known. Use of geochemical and petrographic techniques by Summerfield (1983) enabled him to distinguish between silcretes formed in pan and deep weathering environments, and thus to make palaeoclimatic inferences from the latter group of silcretes. A further problem arises from the fact that it is extremely difficult to observe whether or not a particular duricrust is still forming today.

Thick duricrusts, such as the so-called Kalahari limestone of eastern Namibia, and adjacent areas of Botswana and South Africa form important regional lithostratigraphic marker horizons and may be valuable in establishing regional chronologies (Netterberg 1969, 1978).

The recognition and description of palaeosols, of which duricrusts are effectively a group, may provide valuable palaeoenvironmental data. The existence of a soil horizon within a sedimentary sequence implies that a hiatus in deposition has occurred, with soil formation taking place in a period of geomorphic stability.

Recognition of palaeosols in the sedimentary record may thus provide the basis for regional chronostratigraphic markers and enable indentification of periods of stable environments. From the character of the soil profile, it may be possible to make inferences about palaeoclimates. For example, carbonate rich horizons imply soil formation in a semi-arid to arid climate, whilst organic rich horizons suggest local waterlogging. The degree of soil development may provide some information on the length of time over which pedogenesis has operated.

Although palaeosols have been widely used in studies of palaeoenvironments in North America, Europe, and Australia (Birkeland 1974), there have been few such investigations in Southern Africa (Butzer and Helgren 1972; Helgren and Butzer 1977; Butzer 1984*a*,*b*).

In the south-western Cape, Lambrechts (1983) has argued that the evolution of soils in that region has been strongly influenced by climatic change. Soil characteristics suggest a trend to drier climates and reduced intensity of pedogenic processes during the Late Cainozoic. However, the interpretation of soil characteristics in terms of climates is complex because the nature of soils is influenced by other soil forming factors such as parent material and topography.

3.2.8 Isotopic composition of groundwaters and speleothems

Studies of the chemistry and isotopic composition of groundwater and cave speleothems can provide valuable information on palaeotemperatures and groundwater recharge conditions.

Measurements of the concentrations of inert or relatively inert gases (nitrogen, argon, krypton, and xenon) in groundwater may potentially provide useful information on recharge temperatures. If the age of the groundwater is known through radiocarbon dating, then palaeotemperatures can be determined (Mazor 1972; Heaton 1981). Palaeotemperatures for the Uitenhage aquifer were determined in this way by Heaton *et al.* (1986*a*) and provide a valuable record for this part of Southern Africa which can also be used to calibrate oxygen-18 determinations for other areas, for example, the Cango Cave speleothems.

Variations in the amount of biogenically formed nitrates and 'excess air' in the confined waters of the Stampriet aquifer in Namibia were interpreted by Heaton *et al.* (1983) as a reflection of changes in the environmental conditions during recharge. In part, the nitrate content of the groundwater may be a measure of biological productivity and, hence, rainfall.

The oxygen-18 content of groundwater and deposits precipitated from groundwater, such as speleothems, depends on a number of factors and this makes the interpretation of the oxygen-18 content difficult (Vogel 1983). The oxygen-18/oxygen-16 ratio of groundwater recharged by rainfall depends partly on the humidity conditions over the oceans from which the waters were evaporated. This can be evaluated by comparing the ratios between deuterium and hydrogen and oxygen-18 and oxygen-16 in modern rainwater and the sample under investigation. In addition, the oxygen-18 content of rainwater also depends on the temperature and oxygen-18 content of sea water, which is in turn affected by salinity and global ice volumes. Palaeotemperatures can be determined from measurements of the oxygen-18 content of groundwater and speleothems, provided that assumptions are made about the oxygen-18 content of past ocean waters, and the modern relationships between the oxygen-18 content of rainwater and temperature.

3.3. Biological and archaeological evidence

3.3.1 Pollen

Pollen analysis in Southern and Eastern Africa was pioneered from the 1950s by Van Zinderen Bakker (1951, 1955, 1957, 1962*a*, 1967) and Coetzee (1967). In suitable environments pollen may be preserved for hundreds of millions of years, but most detailed palynological studies are from deposits dating within the last 40 000 years where dating methods are reliable and where differential preservation problems are reduced. The most suitable environments for pollen preservation are acid waterlogged ones, such as peats or beneath vleis and lakes where the sediments have been kept relatively moist, and have accumulated without long depositional breaks. Well-drained alkaline soils are the least suitable. Pollen may be destroyed even in suitable environments, however, by abrasion, oxidation, or bacterial action. The situation of the site is important in assessing the pollen record in terms of past climates. Caves tend to be selective pollen 'traps' and may contain grains of a limited range of wind-blown species or of species brought into the cave by people or animals as bedding or for food. Lakes, pans, or vleis, on the other hand, trap pollen from a wider geographic area although water plants may be over-represented. The nature of the build-up of deposits in springs and peats that may occur in spurts of activity makes them less reliable as accurate records of changes in vegetation through time than cores drawn from lakes or raised peats.

Biases may therefore occur as a result of differences in depositional history, the soil matrix and position of the site, but they are also caused by differences in the types of pollen preserved. Wind pollinated plants such as the Podocarpaceae produce much larger quantities of pollen than insect pollinated ones such as Leguminaceae and tend to be better represented in pollen spectra; also, some pollen is lighter than others and will travel much longer distances. Pollen spectra are therefore not necessarily representative of the composition of a particular vegetation and the tolerances of individual taxa and the modern pollen rain in adjacent vegetation zones (Fig. 5.8, p. 90) need to be studied carefully before palaeoenvironmental interpretations are made (Scott 1982*a*). The absence of a particular species may be as important as its presence in certain areas.

The longest Pleistocene pollen sequences are those from moist organic deposits or springs such as Aliwal North (Coetzee 1967) or Wonderkrater in the Transvaal (Scott 1982*a,b*), or from lakes or vleis such as Groenvlei in the southern Cape (Martin 1968). Cave deposits preserve pollen erratically (Deacon *et al.* 1984*a*) and relatively little work has so far been attempted on offshore Pleistocene pollen from the Southern African coast (Caratini and Tissot 1982). Ideally, it would be useful to be able to correlate pollen sequences in ocean cores with those on land so that the changes can be linked directly to the oxygen isotope chronology.

3.3.2 Charcoal

At places where people have camped and made fires, or where natural fires have left charcoal in the soil, the charcoal can be identified at least to genus and often to species. The species list can give some idea of the range of woody plants and, by inference, environmental conditions in the site surrounds. There is theoretically no limit to the age of identifiable charcoal, but archaeological research has shown that it is only within the last 150 000 years in Southern Africa that people have systematically occupied caves and made fires there. The samples thus far analysed date within the last 70 000 years. Charcoal can be both dated by radiocarbon and identified, thereby increasing the reliability of interpretations of change.

In Southern Africa charcoal has been identified in the southern Cape from Boomplaas and Buffelskloof rock shelters and from the Knysna forest (Deacon *et al.* 1983; Deacon *et al.* 1984*a*), and from Ravenscraig, Colwinton, and Bonawe rock shelters in the north-eastern Cape (Tusenius 1986), from Siphiso shelter in Swaziland (Prior 1984) and from Iron Age sites in Natal (Van Schalkwyk, in Maggs 1982). The longest sequence, that from Boomplaas cave, covers the last 70 000 years and spans the period of the Last Glacial Maximum as well as of the Present Interglacial. Significant changes in the range of woody species selected for firewood can be seen and the climatic interpretation of these changes agrees well with that derived from micromammals at the same site (Deacon *et al.* 1984*a*).

Scholtz (1986) has added to the kinds of palaeoclimatic inferences that are possible from species identifications by introducing a quantitative analysis of the anatomy of the wood preserved as charcoal fragments (xylem analysis). Potentially, these data appear to carry information on both rainfall regime and abundance as well as relative temperature.

Charcoal can be used as direct evidence for vegetation changes, but the potential effects of human selection of firewood from an available supply should be kept in mind. On the other hand, the encouraging agreement between charcoal and micromammalian data at Boomplaas suggests that despite these selection processes, charcoal analysis has considerable potential for palaeovegetation reconstruction.

3.3.3 Tree rings

Several pilot studies have been conducted in Southern Africa to test various of the larger tree taxa for their suitability for both dendrochronology and dendroclimatological reconstruction. In the Transvaal, historic weather records have been compared with tree rings of the last century (Gillooly 1975), and longer climatic records are available from Natal (Hall 1976) and from the southern and south-western Cape (Curtis *et al.* 1979; Dunwiddie and La Marche 1980; Tyson and McNaughton 1980).

The dendroclimatological method involves the measurement of growth rings and the identification of seasonal, annual and unusual patterns of thicker and thinner growth which can be linked to rainfall and/or temperature variability. Where a number of trees are cross-correlated, it is the unusual rings and the patterns of growth increments that are used to link the record of one tree with another. The nature of the climatic signal contained in the pattern of growth can be calculated using a climatic response function based on known rainfall and temperature variations during recent times, and can also provide a fire frequency record (Dunwiddie and La Marche 1980). Pilot studies are essential for the interpretation of these data because different species may respond in different ways to rainfall at particular seasons and the combination of certain temperature anomalies and rainfall are not easily factored out in such cases. Furthermore, the location of individual trees will also influence their sensitivity to rainfall and temperature and this can only be assessed if a large sample is taken. There is nevertheless considerable potential in this method for using tree-ring records as an independent check against early historic data and for discovering long-term variations in local climate.

If long-lived trees can be found in Southern Africa, it is possible potentially to measure oxygen isotope ratios, *D/H* ratios of non-exchangeable hydrogen atoms and stable carbon isotopes in tree rings that would reflect mean annual temperature and humidity (Leavitt and Long 1984; Krishnamurthy and Epstein 1985).

3.3.4 Larger mammals

Faunal assemblages of larger mammal taxa were used until fairly recently primarily for relative dating taking the presence or absence of certain extinct genera and species as an indication of the age of the sample. The first such scheme, which divided the Pleistocene record into four Faunal Spans, was proposed by H. B. S. Cooke (1967), and was modified by Hendey (1974) who proposed that they be termed Land Mammal Ages. Faunal samples dating between 125 000 and 10 000 years BP are grouped within the Florisian Land Mammal Age and they include four species that became extinct between 12 000 and 9500 years ago, namely the giant Cape horse, *Equus capensis*, the giant buffalo, *Pelorovis antiquus*, the giant hartebeest, *Megalotragus priscus*, and a springbok, *Antidorcas australis* (Klein 1980). Within the last 9500 years all assemblages include only species that were extant in historic times, with the exception of *Antidorcas bondi* dated from associated charcoal at Kruger Cave in the Transvaal to 7570±100 BP (Wits-1432) (Brown and Verhagen 1985). Although the extinction of the above species may have been caused partly by climatic change at the end of the Pleistocene and over-predation at this critical time by people with improved hunting technology, the Florisian Land Mammal Age is a chrono-stratigraphic term and has no palaeoclimatic connotations.

Bone is well preserved in most human occupation sites and suitably situated carnivore lairs dating within the last 10 000 years, but is more rarely found in older deposits. Even with a good comparative collection, only a small proportion of the bones (mainly teeth and articular surfaces) can be identified to genus and species because the bones have been broken up to extract marrow and have been weathered or trampled.

Inferences on palaeoenvironments are drawn from the relative proportions of different species and their modern habitat preferences. These proportions are calculated from the minimum number of individuals (MNIs) estimated from the identifiable bones and teeth and are usually expressed as being dominated either by grazing or browsing species indicating a predominantly grassland or predominantly bushy environment respectively. It has been demonstrated that grazer-dominated assemblages in the southern Cape are characteristic of cooler periods when grassland replaced the fynbos vegetation in many areas, whilst browser-dominated faunas characterize interglacials (Klein 1972*a,b*, 1980, 1983). Changes through time at the level of the grazer/browser ratio mentioned above represent a significant alteration in vegetation and therefore in general environmental conditions. This scale of change can be seen not only in fauna at human occupation sites, but also in faunal assemblages of similar age that were accumulated by collectors other than man, for example carnivores and owls (H. J. Deacon 1979; Klein 1980; Deacon *et al.* 1984*a*).

Apart from the relative proportions of grazers and browsers, the mean size of individuals of certain species may also carry palaeoenvironmental information. Klein (1983, 1986) has shown that amongst modern populations of the black-backed jackal, *Canis mesomelas*, individuals in Southern Africa are larger than in Eastern and Central Africa suggesting that size varies along a temperature gradient correlated with latitude (Fig. 6.8), p. 140. In a prehistoric sample from a carnivore lair at Swartklip on the south-western Cape coast, the black-backed jackals are markedly larger than their modern counterparts suggesting that the remains were accumulated at a time of lower temperatures than those of the present. Fluctuations in the mean individual size of small herbivores, on the other hand, appear to vary more with precipitation than with temperature, browser size increasing and grazer size decreasing with increasing rainfall (Klein 1984).

3.3.5 Micromammals

In addition to the bones of larger mammals hunted by people and carnivores, cave and rock shelter sites often preserve the remains of micromammals (mostly small rodents and insectivores with some bats), eaten in most cases by the barn owl (*Tyto alba*), but also by larger species such as the eagle owl. The birds regurgitates the bones after the prey is consumed and these bones represent a particularly useful source of palaeoenvironmental information because the hunting patterns of the owls are different from those of carnivores and people and are subject to different selection processes. The smaller body size of the micromammals make them more habitat specific and therefore a more sensitive indicator of habitat changes than larger mammals. Changes in species composition of micromammalian samples not only pre-date those observed in larger mammal samples at the same site at times of marked climatic change, but also occur at times of more subtle environmental change where no marked differences are discernible in larger mammal samples, although sample size may be a factor contributing to such apparent patterns (Avery 1982*a*; Deacon *et al.* 1984*a*).

Although some studies of micromammals were published in the 1960s on deposits older than the late Pleistocene (De Graaff 1960; Cartmill 1967), the first research projects designed to recover palaeoenvironmental information from Late Pleistocene and Holocene owl pellet remains were undertaken in the 1970s in the Namib (Brain 1974) and in the southern Cape (Avery 1982*a*). They demonstrate that temperature and possibly moisture or rainfall changes can be estimated from both the species composition of samples and from changes in the mean size of individuals in certain species taken from long-sequence sites.

In addition to the study of prehistoric micromammalian samples, the distribution of modern species has been studied as an indication of past climates acting as isolating mechanisms in disjunct distributions. In a study of the distribution and status of the South African shrew, *Myosorex*, Brain and Meester (1964) show that two species were endemic and are separated by over 1600 km from their nearest East and Central African neighbours. This suggests that their present-day distribution is the result of rainfall changes that took place after the formation of the Transvaal australopithecine cave deposits some 1–2 million years ago.

Stuckenberg (1969) undertook a similar study of snakes and frogs in an attempt to demonstrate the effect of reduced temperatures in Southern Africa in the past. Calculating an 'effective temperature' (ET) index, he was able to draw ET isolines to show which regions would be most affected by a reduction in ET in the Last Glacial and suggests that many of the tropical snakes at present found in the Transvaal and Natal would not have been found south of the Limpopo, whilst the ranges of snakes found at present to the south would have extended into Zimbabwe (Stuckenberg 1969, p. 188). His prediction that the combined influence of lower temperatures, lower rainfall and higher wind speeds during the Last Glacial would have created a different biotic pattern from that of the present has been confirmed in studies of fossil micro-mammal and some larger mammal samples. Disjunct distribution of the melanistic lizard *Cordylus cordylus* in the south-western Cape has led Mouton (1986) to suggest that the south-western Cape coastal region was subjected to more adverse climatic conditions than adjacent inland areas during the Last Glacial.

3.3.6 Marine shell

Remains of shellfish eaten by prehistoric people and deposits of shells resulting from natural mortality are known at coastal and inland sites. In terms of palaeoenvironmental data, the changes in proportion of different shellfish species reflect changes in ocean temperature as well as the changing configuration of the shore as rocks or sandy beaches are exposed. In addition, the isotopic composition of shell can also be used as an indication of ocean temperature and the extent of the Antarctic ice sheet (Shackleton 1982). However, because of changes in sea level (which dropped as much as 140 m at times of glacial advance) only those shell middens dating to interglacial periods can be studied because the rest are now covered by the sea. Because shellfish eating appears to have become common practice only since the Last Interglacial, this method is limited to deposits dating within the last 125 000 years. The technique has so far been used at one site in Southern Africa, at Klasies River Mouth on the southern Cape coast, where the oxygen isotope composition of shell indicates a late Last Interglacial and early Last Glacial age (Shackleton 1982).

3.3.7 Stable carbon and nitrogen isotopes in bone

The stable isotopes carbon-12 and carbon-13 have been measured according to a PDB standard in animal and human bones using a mass spectrometer and they show a correlation with plant foods eaten at the base of the food chain (Vogel 1978). Fossilized bone in which the bone collagen has been lost cannot be used in this type of analysis and the method has been limited so far to more recent bone samples dating within the last 100 000 years or less.

All green plants assimilate carbon dioxide from the atmosphere using a photosynthetic pathway and three species-specific pathways are known, namely C_3, C_4 and CAM. The pathways are identifiable from the carbon-12/carbon-13 ratio. In the process of photosynthesis the atmospheric isotope ratio is depleted and Fig. 3.1. illustrates this fractionation from C_3 and C_4 photosynthesis at different metabolic stages. CAM plants use both pathways, one by day and the other by night, but as they include a

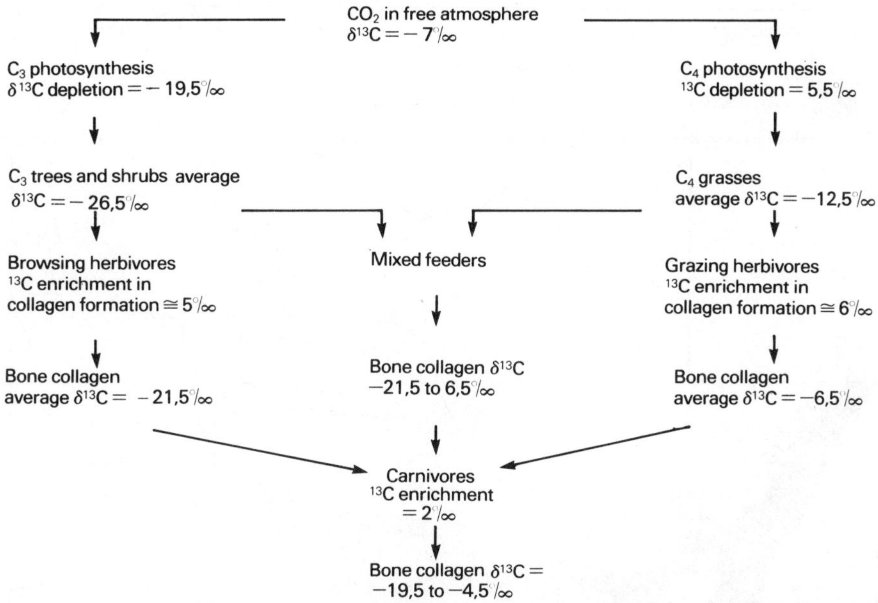

Fig. 3.1 Carbon pathways in a typical African savannah with C_4 grasses, and C_3 trees and shrubs, showing fractionation of the carbon-13/carbon-12 ratios at different metabolic stages (after Van der Merwe 1986). (Reprinted from: *Palaeoecology of Africa*, (ed. E. M. van Zinderen Bakker, J. A. Coetzee, and L. Scott) Volume 17. Southern African Society for Quaternary Research, 1986. A. A. Balkema, P.O. Box 1675, Rotterdam, Netherlands.)

relatively small range of plants (cacti and some euphorbias) that are not particularly important in the diet of most herbivores, they have not been considered further.

It is more difficult to reconstruct the diet of carnivores and omnivores (such as man) from the carbon isotope ratio of their bone collagen not only because of the greater variety of their foods, but because of additional small fractionations that occur during the formation of different animal tissues that are at present not fully understood (Van der Merwe and Vogel 1983; Sealy and Van der Merwe 1986). Nevertheless, this has proven possible where there is good evidence for the range of foods consumed. In Southern and Eastern Africa pastoralists, hunter-gatherers and agriculturalists can be differentiated (Ambrose and De Niro 1986).

In South Africa the percentage frequencies of C_3 and C_4 plants are significantly different in winter, summer and all-year rainfall regions (Fig. 3.2). This has palaeoclimatic implications because low temperatures during the growing season (i.e. season of maximum rain) favour C_3 species and high temperatures during the growing season favour C_4 species. Thus, palaeotemperature changes can be discerned in marginal

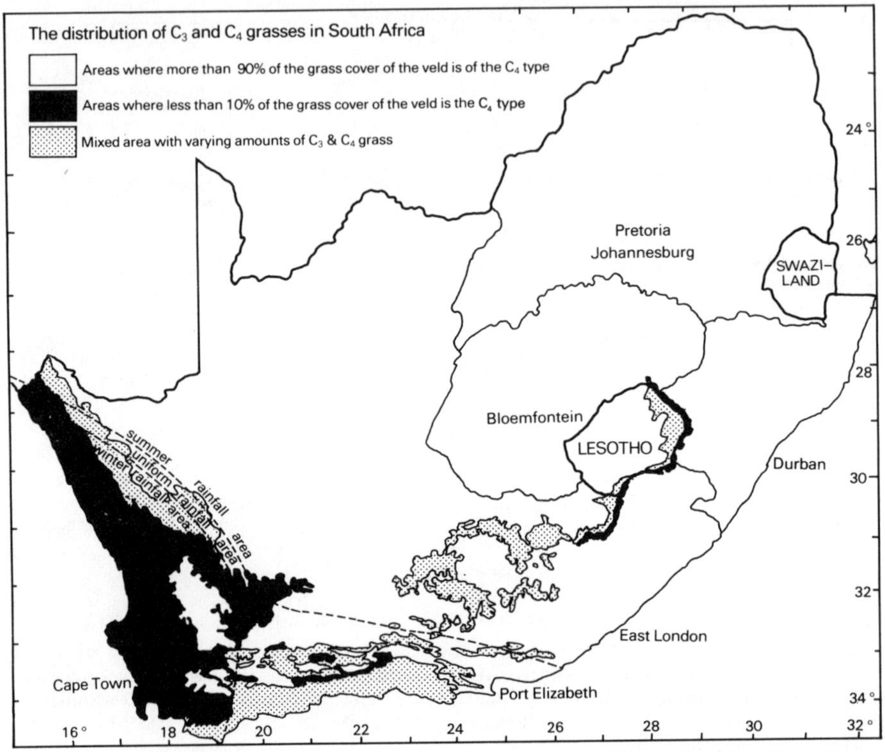

Fig. 3.2 The distribution of C_3 and C_4 grasses in South Africa (after Vogel 1978).

areas where the relative frequencies of C_3 and C_4 plants have changed with marked and sustained shifts in mean annual temperature. Changes in the season of rainfall may also be discerned from the relative proportion of C_3 and C_4 plants. The best samples to test for palaeotemperature and rainfall changes are those from sites with long sedimentary sequences near the boundaries of C_3 and C_4 regions, or from mountainous areas where altitude plays an important part in determining the annual range of temperature (Vogel 1983).

Heaton *et al.* (1986*b*) have shown that the nitrogen-15/nitrogen-14 ratio in bone is also an indicator of diet, and has particular significance for palaeoclimatic studies because modern samples show a correlation between nitrogen isotope ratios and rainfall (Fig. 3.3). This method could be used to indicate major changes in precipitation through time.

3.3.8 Archaeological proxy evidence

The use of archaeological proxy evidence in palaeoclimatic reconstruction is based on the assumptions that (a) the density of archaeological

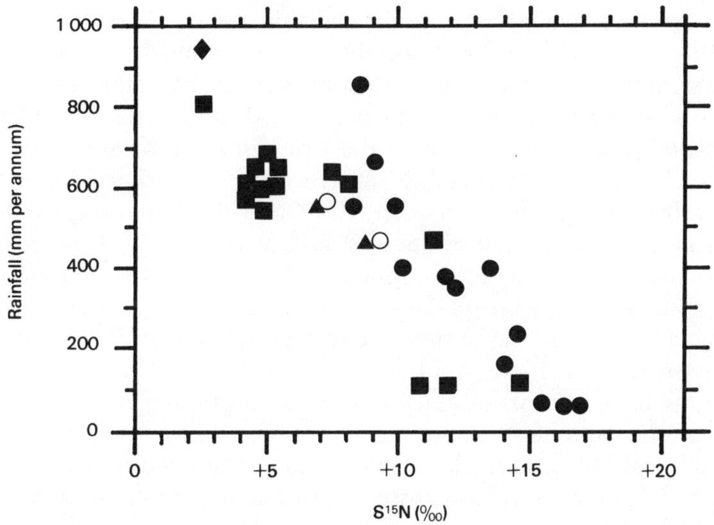

Fig. 3.3 Nitrogen isotope ratios of human (solid circles), elephant (squares), wildebeest (open circles), giraffe (triangles), and zebra (diamonds) bones plotted against the general range of mean annual rainfall for the collection areas in Southern Africa (after Heaton *et al.* 1986*b*).

sites and their location is a direct (if gross) reflection of the number of people in an area over a particular period of time, and (b) that this density will be related to the productivity of the environment in that area. Implicit, too, is the assumption that people moved or changed their hunting and gathering methods only as the result of environmental change. There are few modern analogues for such behaviour, and there is the danger that such proxy data could lead to a circular argument.

The importance of understanding geographical and climatic variables in accounting for the distribution of Stone Age industries was stressed as early as 1881 (Gooch 1881) and again when the first comprehensive survey of South African sites was made (Goodwin and Van Riet Lowe 1929). It was also appreciated that demographic change could be expected to have occurred when climates changed—for example, arid areas would be largely evacuated in times of lower rainfall (Van Riet Lowe 1930, 1937; Leakey 1931, p. 240; Smuts 1932, p. 126). During the 1950s more detailed distribution maps of Stone Age and Iron Age sites were compiled by Clark (1957) and Summers (1960) for Zambia and Zimbabwe, respectively and the results demonstrated both that some artefact types were more common in certain ecological regions than in others, and that variables such as the distribution of tsetse fly and sleeping sickness were negatively correlated with the distribution of Iron Age sites.

The increasing availability of radiocarbon dates, particularly for the last 20 000 years, has made it possible to map prehistoric site distributions in a more systematic way. In mapping the distribution of dated Later Stone Age sites in 1000-year periods within the last 10 000 years, for example, it was shown that in the drier interior of South Africa in regions of low and unpredictable rainfall at present, there are very few dates for the time period between 9000 and 4500 BP suggesting that population density at that time was relatively low (J. Deacon 1974, 1984). Beaumont (1986) has mapped dated sites in the northern Cape and has concluded that Griqualand West was under-populated during the Last Glacial, but at the same time the numerous pans in Bushmanland to the west provided a focus for human occupation. He suggests that the two regions have complementary culture stratigraphic records implying that when the summer rainfall area of Griqualand West experienced sustained drought conditions, the winter rainfall in Bushmanland increased and made the region more favourable for human settlement.

On a broader time scale, gross changes have been documented over the last 125 000 years (Deacon and Thackeray 1984) that show a low visibility of Stone Age populations during much of the Last Glacial cycle and a relatively high number of dated occurrences in the Last and Present Interglacials. It would seem that people found it easier to cope in regions of steep ecological gradients, where there is a closer nesting of ecological zones, than in areas like the Karoo where the environment is more uniform over a larger area and where they would be more vulnerable to long-term temperature or precipitation changes. The data argue against hypotheses such as that by Lee (1963) which assume that population densities and distributions were stable over long periods of time during the Pleistocene. They demonstrate instead that people were indeed affected by changes in temperature and rainfall which took place over the last 125 000 years and more, and that in some areas there were long periods of time (tens of thousands of years) when population densities were considerably lower than at other times.

4 Palaeoenvironments in the desert and semi-desert ecozones

4.1 Introduction

In reviewing the evidence for Late Quaternary palaeoenvironments in Southern Africa we have chosen to discuss the data first by rainfall regime (desert, summer, winter, and all-year rainfall regions) and second by ecozone. The ecozone subdivisions have been used by Klein (1980, 1984) and are adopted here because they take account not only of climate, but also of vegetation and fauna, and are therefore more meaningful in palaeoenvironmental reconstructions. The geographic extent of the ecozones is summarized in Fig. 4.1. Radiocarbon dates mentioned in the text are listed in full in Appendix 1.

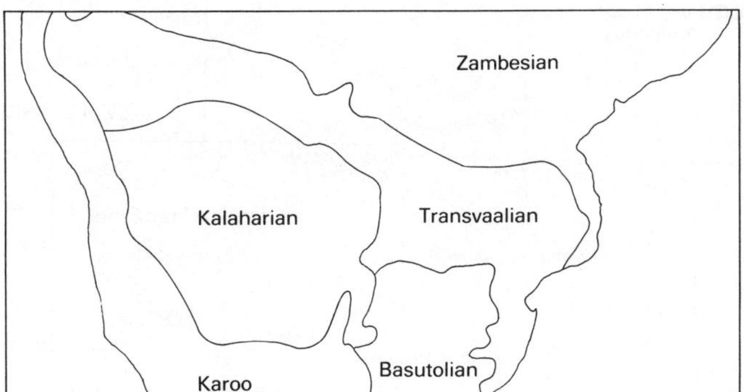

Fig. 4.1 Southern African ecozones (after Klein 1980).

4.2. The Namib

The Namib, with its cool hyperarid to arid climate and unique biota, is sufficiently distinctive to be treated as a separate region within the Karoo–Namaqualian Ecozone (Figs. 4.1 and 4.2). It is the driest region of Southern Africa and has experienced a low amplitude of climatic changes in the Late Pleistocene (Ward *et al.* 1983; Lancaster 1984*b*). It extends for 2000 km along the Atlantic coast of Southern Africa between the Olifants River in South Africa at 32° south to San Nicolau in Angola at 14.30° south. The inland margin of the Namib lies at the base of the Great Escarpment, 100–200 km from the coast. The Benguela Current runs parallel to the coast and strong upwelling of cold sub-Antarctic waters occurs, particularly between latitudes 26 and 28° south.

The Namib can be divided into the southern or transitional Namib, a region of rocky plains and low hills with areas of sand sheets and dunes;

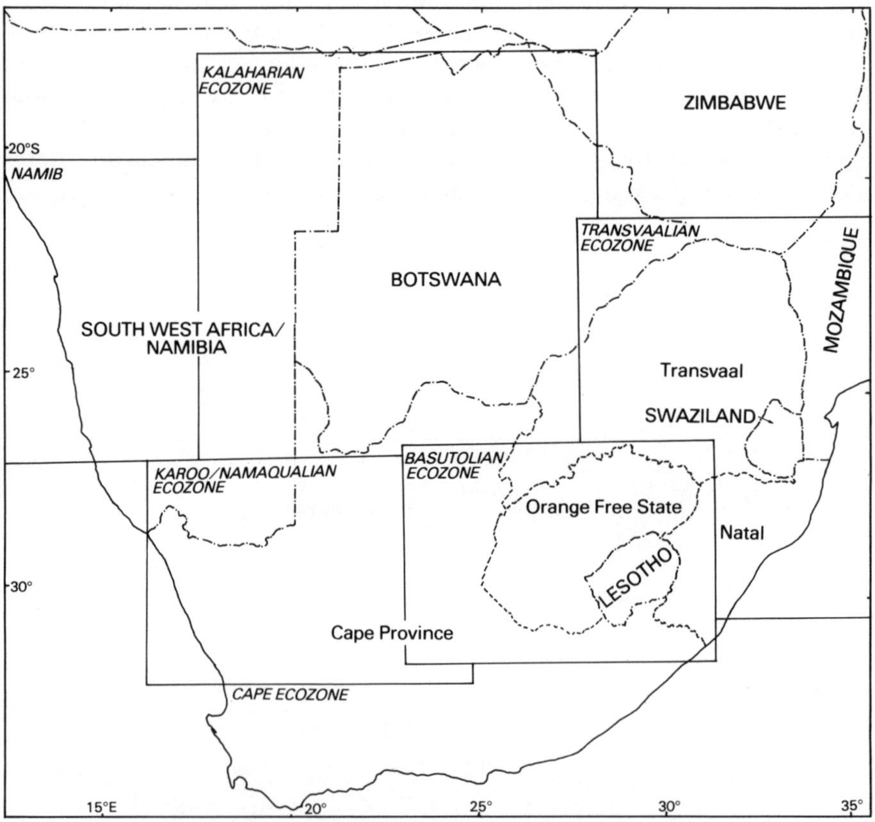

Fig. 4.2 Key to regional ecozone maps in Chapters 4, 5, and 6.

the Namib Sand Sea (Fig. 4.3) between Luderitz and the Kuiseb River; the central Namib plains with scattered ranges of hills and inselbergs which continue north to Damaraland; and the northern Namib where areas of dunes occur along the coast. The climate of the Namib is hyperarid to arid with a steep climatic gradient inland (Seely 1978; Lancaster *et al.* 1984) from the cool foggy coastal zone which receives less than 20 mm of rain annually to the hotter, but less humid areas inland, which receive a scant summer rainfall of 50–100 mm. In the south, the Namib becomes more mesic and grades into the winter rainfall zone of the Cape Province. Throughout the region, vegetation is very sparse. Trees are confined almost entirely to the larger river courses. Grasses spring up rapidly after rains (Seely and Louw 1980), and

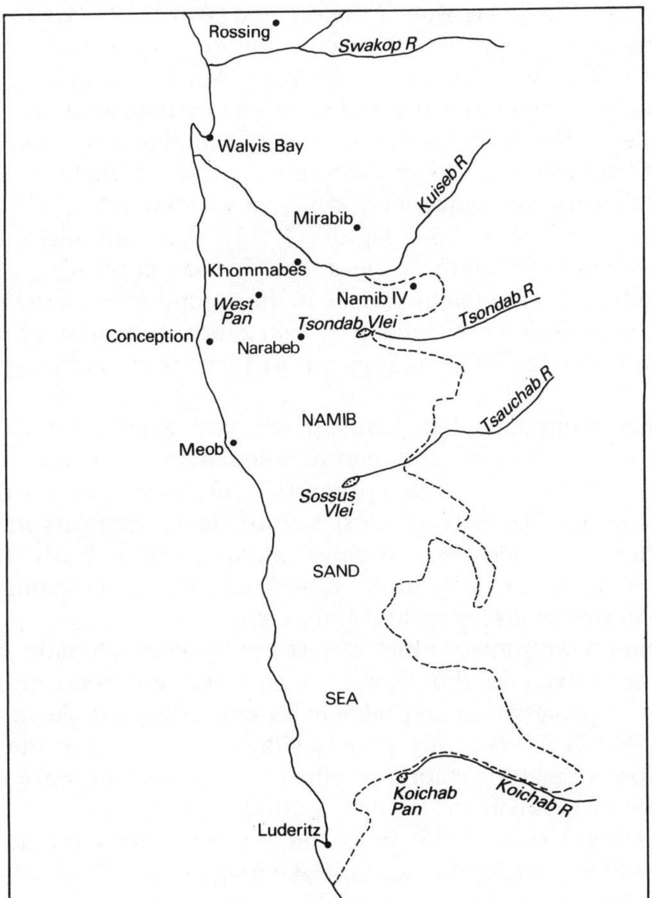

Fig. 4.3 Location of sites in the Namib.

provide grazing for herds of gemsbok (oryx) and springbok, with zebra in the east of the desert.

The causes and age of aridity in the Namib, and their effects upon the evolution of its unique endemic biota have been the subject of considerable debate (Van Zinderen Bakker 1975; Tankard and Rogers 1978; Ward *et al.* 1983). The available geological evidence indicates a long history of arid climates in the region, dating back to the early-mid Tertiary (Van Zinderen Bakker 1975; Ward *et al.* 1983) with hyperarid conditions developed from the late Miocene (7–10 million years ago) onwards (Siesser 1980), following the beginning of strong upwelling by the Benguela Current. Throughout the Quaternary, climatic fluctuations appear to have been of a low amplitude and superimposed upon an arid to hyperarid mean (Lancaster 1984*b*). The surface survival of end-Miocene calcrete palaeosols (Yaalon and Ward 1982) and lacustrine carbonates (Selby *et al.* 1979) suggest that no very humid climates have affected the region in the Late Cainozoic. Ward *et al.* (1983) argue that the Namib has experienced no climate wetter than semi-arid since the end-Miocene, 5 million years ago. Pollen samples from deep sea cores taken on the Walvis Ridge show no significant changes throughout undated Pleistocene sequences, which are dominated by Gramineae, indicating to Caratini and Tissot (1982) that, although there were intervals when the climate was more humid than at present, it was never wet enough to cause a major change in the composition of the vegetation. Similar conclusions were reached by Van Zinderen Bakker (1984) on the basis of a more detailed investigation of Pleistocene pollen from DSDP cores.

Evidence from fluvial deposits (Korn and Martin 1957; Lancaster 1984*b*; Ward 1984*a,b*) and marine sequences (Tankard and Rogers 1978) suggests that there was a progressive increase in aridity throughout the Pleistocene. The average clast size of fluvial deposits in the major river valleys declines from cobbles and gravels in Early Pleistocene deposits to sands and silts in the Late Pleistocene, suggesting a parallel decrease in stream discharge and competence.

In common with many other desert areas, palaeoclimatic evidence is variably preserved in the Namib. There are few depositional basins onshore and preservation of pollen in sediments is poor. As in the Sahara (Rognon 1982) the best preserved evidence comes from the sand seas which show a greater reaction to climatic changes and more contrast in environments between wet and dry periods compared to adjacent gravel plains. As Ward *et al.* (1983) point out, interpretations of palaeoclimates in the Namib must take into account the long narrow shape of the desert, its steep climatic gradient, and the linear oases which penetrate far into the desert along the major river valleys.

4.2.1 Interdune pond and lacustrine deposits

Small areas of mostly thin (<4 m) lacustrine carbonates and associated calcified reeds and gastropod shells are scattered throughout the northern parts of the Namib Sand Sea (Fig. 4.4.) where they are exposed in interdune areas between 50–100 m high linear and star dunes. The deposits, which were mapped as the Khommabes Carbonate Member of the Sossus Sand Formation by Ward (1984*a*), appear to have been laid down at intervals during the accumulation of the Namib Sand Sea.

Sediments of the Khommabes Carbonate Member occur in shallow basins developed in the Tertiary Tsondab Sandstone Formation and overlying Late Tertiary–Early Pleistocene Karpfenkliff conglomerates and gravels of the proto-Kuiseb River (Ward 1984*a*). The pond and lacustrine deposits often display a characteristic association of facies, with white powdery lacustrine carbonates in the centre of the basin, and case hardened deposits and calcite cemented reed stems and organic debris overlying bleached and mottled aeolian dune and interdune sands on the margins (Teller and Lancaster 1986*a*).

Stone Age artefacts of Earlier to Later Stone Age affinities occur in association with the deposits (Shackley 1985) and indicate a long history of episodic and localized ponding of surface runoff and groundwater seepage. The occurrence of artefacts of different ages (especially Earlier

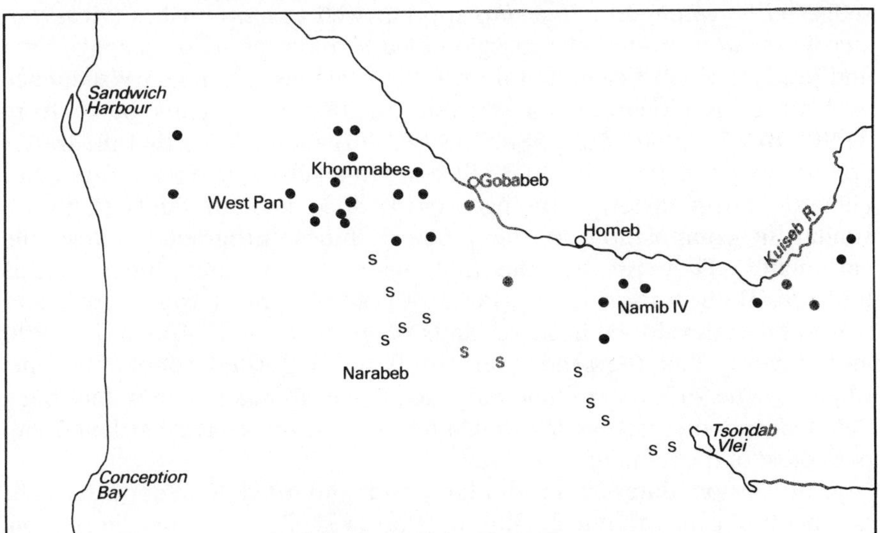

Fig. 4.4 Exposures of interdune pond and marsh deposits in the northern part of the Namib Sand Sea (after Ward 1984*a*).

and Middle Stone Age) at the same site implies reoccupation at intervals when moisture was available within the now arid sand sea. The oldest deposits occur at Namib IV where bones of *Elephas recki* indicate an age of between 700 000 and 400 000 BP (Shackley 1980).

Snails from lacustrine deposits at West Pan include *Melanoides tuberculata, Tomichia ventricosa, Biomphalaria pfeifferei* and *Bulinus* sp. which indicate that fresh to brackish water bodies occurred in interdune areas some 27 500 BP (Teller and Lancaster 1986*a*).

Deposits at Khommabes, a shallow basin situated in an interdune area between 60–100 m high linear dunes 1 km south of the Kuiseb River, were mapped by Ward (1984*a*) and studied in detail by Teller and Lancaster (1985). They record the interaction of fluvial, aeolian, and lacustrine processes at the northern margin of the Namib Sand Sea. The earliest fluvial deposits (Units B and C) consist of sands and silty sands with a high percentage of biotite mica which were deposited in a period of wetter climates by a more extensive flood plain of the Kuiseb River, possibly before the large south–north trending linear dunes has reached this area. They were later cemented by pedogenic calcretes which were subsequently altered to dolocrete. Following this period, the aeolian sands of Unit D, possibly derived by deflation of the flood plain, were deposited on the basin margins. Renewed wetter conditions, or percolation of water from the Kuiseb River, resulted in the growth of reeds and sedges (*Phragmites* and *Juncellus* spp.) as well as nara bushes (*Acanthosicyos horrida*) around the margins of a shallow pond or seepage area and precipitation of calcrete on or in the previous sedimentary sequence as Unit E. Radiocarbon dates from calcified reed stems range from 31 900 to 27 400 BP (Pta-2588, 2589, 2590) and indicate that the wetter conditions prevailed prior to 32 000 BP. In a subsequent period of drier climates, north-easterly dipping cross-bedded aeolian sands (Unit F), similar in composition to the modern dunes, prograded across the carbonates. They provide the first evidence for sand dunes in this location. Radiocarbon dates (Pta-1091, 2584, 2604) from roots, and pedotubules developed in these sands range between 22 400 and 20 900 BP indicating that the sands were probably deposited about this time. Much of the original sedimentary sequence at Khommabes has been deflated and the carbonates of Unit E now form a case-hardened cap over most of the remaining sediments.

Radiocarbon dates from similar pond and reed bed deposits near Gobabeb cluster around 21 500 BP (Pta-2651, 2652) whilst those from Koichab Pan on the southern margins of the sand sea have dates of 23 800–23 300 BP (Pta-3184, 3185). At Kannikwa, in the southern Namib near Port Nolloth, a peat lens from a pan has a radiocarbon date of 23 200 ± 300 BP (Pta-3978) indicating cool and wet conditions at this time (Beaumont 1986).

Calcified reed beds and organic materials also occur along the Namib coast, adjacent to sites at which vegetation growth today is promoted by seepage of groundwater where the contact between dune sand and the underlying Precambrian bedrock comes to the surface. Calcified organic deposits and *Phragmites* reeds with radiocarbon dates ranging between 12 600±140 and 11 700±120 BP (Pta-1238, 1647, 1830, 1831) occur at Conception and Meob, and indicate that increased groundwater discharge occurred prior to 12 600–11 700 BP, when desiccation and calcification of the reeds took place (Vogel and Visser 1981).

A 36-m thick sequence of calcareous mudstones and interbedded sands at Narabeb, some 40 km south of the Kuiseb River, occurs in a similar interdune situation to deposits of the Khommabes Carbonate Member, but provides a very different record of past climates. The deposits were first described by Seely and Sandelowsky (1974) who suggested that they had been deposited at the former end point of the Tsondab River, which now terminates 40 km to the east at Tsondab Vlei. Selby *et al.* (1979) published uranium-234/thorium-230 dates from a lower member of the deposits, which they regard as having been laid down in a short-lived interdune pond. Six calcareous mudstone units were identified by Teller and Lancaster (1986*b*). They are invariably laminated and sandy, and contain halite filled fractures and desiccation cracks. Unit IV contains planktonic diatoms, mainly *Tabellaria fenestrata*, which are characteristic of eutrophic freshwater lakes. The sands are slightly calcareous, but are otherwise similar to those of the Tsondab Sandstone Formation, which underlies the deposits, as well as to the adjacent dunes. Some sandy units contain chips of calcareous muds deflated from nearby lake beds. Teller and Lancaster (1986*b*) argue that the deposits represent an alternation of wet and dry periods. The calcareous mudstones were deposited in a shallow lake at a former end point of the Tsondab River, which was dammed by dunes. The sands represent aeolian deposition in intervening dry periods. Desiccation is evidenced by mud cracks and halite in the upper parts of the mudstone units. The age of the Narabeb deposits is uncertain. Four radiocarbon dates on the mudstones range betwen 26 000±400 and 20 320±300 BP (Pta-3704, 3759, Beta-9115, 9116), but the uranium-234/thorium-230 dates obtained by Selby *et al.* (1979) from what appears to be the basal mudstone suggest an age of more than 200 000 years for the whole sequence. The association of archaeological materials of Earlier and Middle Stone Age affinities with the deposits (Shackley 1985) indicates that they are more than 25 000 years old.

The present end point of the Tsondab River is Tsondab Vlei, to which the river floods ephemerally. The penetration of the Tsondab River to Narabeb clearly indicates increased flooding and greater discharge as a result of higher rainfall in the highland parts of its catchment (Lancaster 1984*a*) in the period of 30–20 000 years ago, as well as the absence of dunes east of Narabeb. By 14 000 BP its terminus had shifted eastward to a point 5–10 km west of Tsondab Vlei, as evidenced by radiocarbon dates of 14 300±130 and 13 300±90 BP (Pta-1043, 1502) on gastropods from 2–3 m thick silt deposits (Vogel and Visser 1981). Since then, the course of the Tsondab River has been blocked by several 50–100 m high dunes and Holocene sedimentation has been confined to the vicinity of the present vlei (Lancaster 1984*a*).

4.2.2 Fluvial deposits

The Namib Desert is crossed by a number of rivers draining from the highland areas to the east. Apart from the perennial Kunene and Orange rivers, both of which rise well outside the arid zone, most flow ephemerally. From the Kuiseb northwards, most of the large rivers have been known to reach the Atlantic on occasions. Between the Kuiseb and the Orange, no streams reach the sea. This is the result of the increasing aridity of the highland catchment areas and the presence of the dunes of the Namib Sand Sea. The Tsondab, Tsauchab, and Tsams rivers penetrate up to 60 km into the sand sea, to end against the dunes in extensive playas. Most of the Namib rivers have deposits in their valleys which record a sequence of aggradation and incision spanning much of the Late Cainozoic. The importance of the fluvial sequences as indicators of past climates was noted by earlier workers like Gevers (1936), Mabbutt (1952), and Korn and Martin (1957). Subsequent work by Rust and Wienecke (1974, 1980), Marker (1977*a,b*), Ollier (1977), Ward (1982, 1984*a,b*), Lancaster (1984*a*), and Heine (1985) has resulted in the elucidation of the stratigraphy of these sequences, especially for the Kuiseb, but failed to establish whether they reflect climatic or tectonic changes in the Namib and its hinterland. However, it is clear that all the fluvial deposits were laid down in arid to semi-arid climates (Ward *et al.* 1983). Most of the deposits are of early to mid-Pleistocene age (Ward 1984*b*), but in the Kuiseb valley 30 m of micaceous silts (Marker and Muller 1978) accumulated in the period 23 000–19 000 BP (Vogel 1982) in conditions of low stream discharge and competence. The deposits were laid down on a low energy flood plain in a section of the valley upstream from Gobabeb (Ward 1984*b*), rather than at the end point of the river (Vogel 1982) or behind a dune dam (Rust and Wienecke

1974), as evidenced by the abundance of cross-bedded sands of fluvial and derived aeolian origins (Ward 1984*b*). They indicate low stream energy and less intensive floods suggesting drier conditions.

4.2.3 Speleothems from Rössing Mountain

Heine and Geyh (1984) report radiocarbon dates on speleothems from a small limestone cave near Rössing Mountain, 35 km inland from Swakopmund on the hyperarid Namib coast. They occur in an area where rainfall is currently less than 20 mm per year. According to Heine and Geyh (1984) the development of speleothems in this area indicates increased rainfall and run-off in the immediate catchment area of the cave. A series of ten radiocarbon dates indicates that the most recent speleothems were deposited in a period of more humid climates between 41 500 and 26 500 BP.

4.2.4 Micromammals

A sample of micromammalian bones that dates within the last 6000 years from Mirabib rock shelter in the central Namib plains was studied by Brain and Brain (1977). Using the presence of the golden mole (*Eremitalpa granti namibensis*) as an indicator of sandy habitats, it appears that the Kuiseb River, 24 km to the south, has remained a barrier to dune sand during the last 6000 years. The presence of *Malacothrix typica* and a higher relative frequency of gerbils in all but the uppermost units suggest that a more favourable habitat, with moister conditions and more grass cover, was present in the area until recently. Drier conditions than those in the area today have occurred twice in the last 500 years, whilst a period with climates similar to today was centred around 5200 BP.

4.2.5 Stable carbon isotopes

In contrast to the changes observed in the carbon-13 content of bone collagen in zebra teeth from Melikane Cave in Lesotho, zebra teeth from Late Pleistocene (*c.* 70 000–20 000 BP) and Holocene (7000 BP) units at Apollo 11 Cave excavated by Wendt (1972) on the eastern margin of the southern Namib Desert show no appreciable difference (Table 4.1). This led Vogel (1983) to suggest that C_3 grasses did not spread into southern Namibia during the Late Pleistocene, indicating that winter rainfall did not extend much further north in the Late Pleistocene than it does today.

Table 4.1. Carbon-13 content of collagen from Equus teeth excavated from Apollo 11 Cave in southern South-West Africa/Namibia by Wendt (1972). The last column lists the contribution of C_3 (winter rainfall) plants to the diet of the animals as deduced from the isotopic composition of the teeth. The results show no appreciable difference between the Holocene and Late Pleistocene samples (from Vogel 1983).

Sample	Approx age (years BP)	Carbon-13 content	% C_3 in diet
B156/Z1	Recent	−12.6	36
B157/Z2	7000	−14.1	47
B158/Z3	20 000	−14.9	53
B159/Z4	70 000	−14.8	53

4.3 Chronological summary

Presently available dated palaeoclimatic information for this ecozone is confined to the last 32 000 years. In the period from prior to 32 000 to 20 000 BP the region appears to have been substantially wetter than today with interdune lake and pond deposits in the Namib Sand Sea. River flow was also stronger than at present. A drying trend was present locally after 27 000–26 000 BP. Cool conditions are evident around 23 000 BP. The period spanning the Last Glacial Maximum (20 000–16 000 BP) appears to have been dry in the Namib. The interval from 16 000 to 12 000 BP sees a renewal of increased moisture availability in the Namib, continuing until around 11 000 BP. For the Holocene, the climate appears to have been slightly moister than it is today, with a dry interval around 2500 BP.

4.4 The Karoo and Bushmanland

The Karoo and Bushmanland also form part of the Karoo–Namaqualian Ecozone. Until very recently, virtually no palaeoclimatic evidence was recorded from this part of Southern Africa, although the widespread occurrence of diatomite deposits in pans was noted by Kent (1947). Recent investigations of pan deposits and archaeological surveys have indicated that a wealth of palaeoclimatic evidence exists in this region and may prove valuable in the correlation of sequences and events in the summer and winter rainfall areas of the subcontinent.

The Karoo and adjacent areas of Bushmanland (Fig. 4.5) are characterized by extensive rock plains at an altitude of 900–1200 m developed on sandstones and shales of the Karoo System. Mesas, often capped by dolerite sills, occur at intervals. Towards the western and northern margins of this region Precambrian rocks crop out and form low ranges of hills. In the southern part of the Karoo, proximity to the Cape Fold Belt imparts a strong west to east orientation to structures and drainage. A large area of endoreic drainage and pans occurs in the western Karoo.

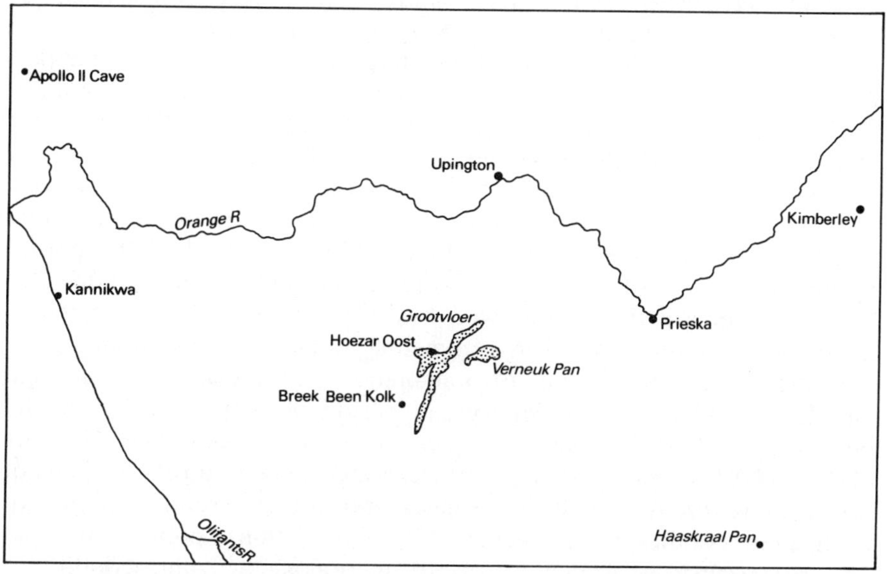

Fig. 4.5 Location of sites in the Karoo–Namaqualian Ecozone.

The Karoo experiences a semi-arid to arid climate with an annual rainfall of 150–400 mm. This falls mostly in the summer in the wetter north and east of the region. The drier western and southern Karoo receive rainfall during the winter or at the equinoxes. Vegetation in the region is distinctive and is characterized by low shrubs and succulents with scattered grass, which becomes abundant after good rains. Trees are confined to the larger watercourses.

4.4.1 Pans and palaeolakes

Pans and shallow endoreic drainage lines, known locally as vloers, constitute the only depositional basins in this region. Stratigraphic and sedimentological investigations of deposits at Haaskraal Pan reported by Partridge and Dalbey (1986) provide evidence of palaeoclimatic changes in the central Karoo since the mid-Pleistocene. Deposits preserved in the pan complex include alluvial gravels and sands deposited as the braided channel of the Zoetvlei River crossed the pan in response to changes in discharge and meander geometry. The large clast size of the gravels indicates that stream competence was much greater than now, implying increased run-off and wetter climates. Loams and clayey loams accumulated as overbank deposits in the distal parts of the river flood-plain, whilst vlei and pan deposits were laid down within abandoned channels. Yellowish red coarse sandy loams were transported as colluvium from adjacent hillslopes. Varying degrees of post-depositional pedogenesis and calcrete formation affected most of the sedimentary units. The resulting series of palaeosols is unparalleled in Southern Africa and records a series of episodes of desiccation of the pan in dry climates, during which archaeological visibility increased as scattered populations regrouped around water sources. Major Late Pleistocene periods of desiccation and calcrete formation occurred in Middle Stone Age times and again between 28 000 and 18 000 BP. In the latter period approximately 2 m of pan deposits were deflated. The periods of desiccation were separated and followed by more humid climates in which deposition took place in vlei environments, with local colluviation.

In the western Karoo and Bushmanland there is evidence from high lake levels and pans of a widespread period of wetter climates in the period 30 000–14 000 BP. At Breek Been Kolk, Kent and Gribnitz (1985), and Beaumont (1986) have described thick deposits of shells of *Tomichia ventricosa* which they suggest indicate an extended period of freshwater conditions. However, Grindley (1986) points out that *Tomichia ventricosa* frequently occurs in brakish or saline conditions. Radiocarbon dates on the sediments (Beaumont 1986) range from a basal 38 000±1700 BP (Pta-3958) to 13 900 ±130 BP (Pta-1760) at the top of the deposit, implying a long history of shallow and possibly seasonal lacustrine conditions at this pan. In the Holocene, desiccation occurred prior to 8000 BP. At Hoezar Oost on Grootvloer, a palaeolake with a depth of 2–4 m is evidenced by a nearshore bar thinly covered with shells of the freshwater mollusc *Limnia caffra* with a radiocarbon date of 17 800±190 BP (Pta-3894) (Beaumont 1986).

4.4.2 Archaeological proxy evidence

Beaumont (1986) has discussed the implications of changes in archaeological visibility in Bushmanland and Griqualand West in the Late Pleistocene and Holocene. Archaeological visibility was high in both areas during the period of the Last Glacial Maximum, suggesting that they offered favourable habitats for Later Stone Age hunter gatherers.

This pattern is in contrast to that indicated for Southern Africa as a whole by Deacon and Thackeray (1984).

4.5 Chronological summary

The small amount of dated palaeoclimatic evidence available for this region suggests that generally wetter conditions prevailed during the period from 30 000 to 13 000 BP, but it is not possible to indicate whether this was summer or winter precipitation.

5 Palaeoenvironments in the summer rainfall ecozones

5.1 The Kalaharian Ecozone

The Kalahari is part of an extensive sand covered plain at an elevation of 800–1200 m which extends from the Orange River to southern Angola (Fig. 5.1). To the east it is bounded by the highlands of eastern Botswana and western Zimbabwe, whilst to the west are the mountains of central Namibia. Most of the drainage in the region is internal and pans are common in southern and south-western parts of the Kalahari. The Okavango, Cuando, and Cuito rivers rise in the humid highlands of Angola, and flow south-west to form large inland deltas and swamps in north-western Botswana, and adjacent areas of Namibia and Zambia, where much of their discharge is lost by evaporation and evapotranspiration. Some water spills over from the Okavango delta and finds its way to the Makgadikgadi pans.

Although often called a desert, the Kalahari is more correctly termed a thirstland or edaphic desert. There is no permanent water outside the swamp zone, but the rainfall of 500–800 mm in northern areas, declining to 150–200 mm in the south-west, supports a cover of tree and bush savannah. Even today, the Kalahari supports large numbers of game animals, including elephant and buffalo in the better watered northern areas. There are extensive herds of springbok, blue wildebeest, hartebeest, and Burchell's zebra throughout much of the region.

Since the Cretaceous, the Kalahari has acted as a major sedimentary basin (Tankard *et al.* 1982) and contains up to 500 m of terrestrial sediments, mainly sands and marls, with occasional playa sediments which appear to have accumulated in mostly semi-arid to arid environments (Du Toit 1954; Boocock and Van Straten 1962; Wright 1978; SACS 1980). Many of the sediments, termed the Kalahari Group by SACS (1980) have been post-depositionally calcified or silicified (Watts 1980; Summerfield 1983). They are capped by the surface Kalahari sands, which are largely aeolian in character (Baillieul 1975) and mantle an area of up to 2.5 million km^2.

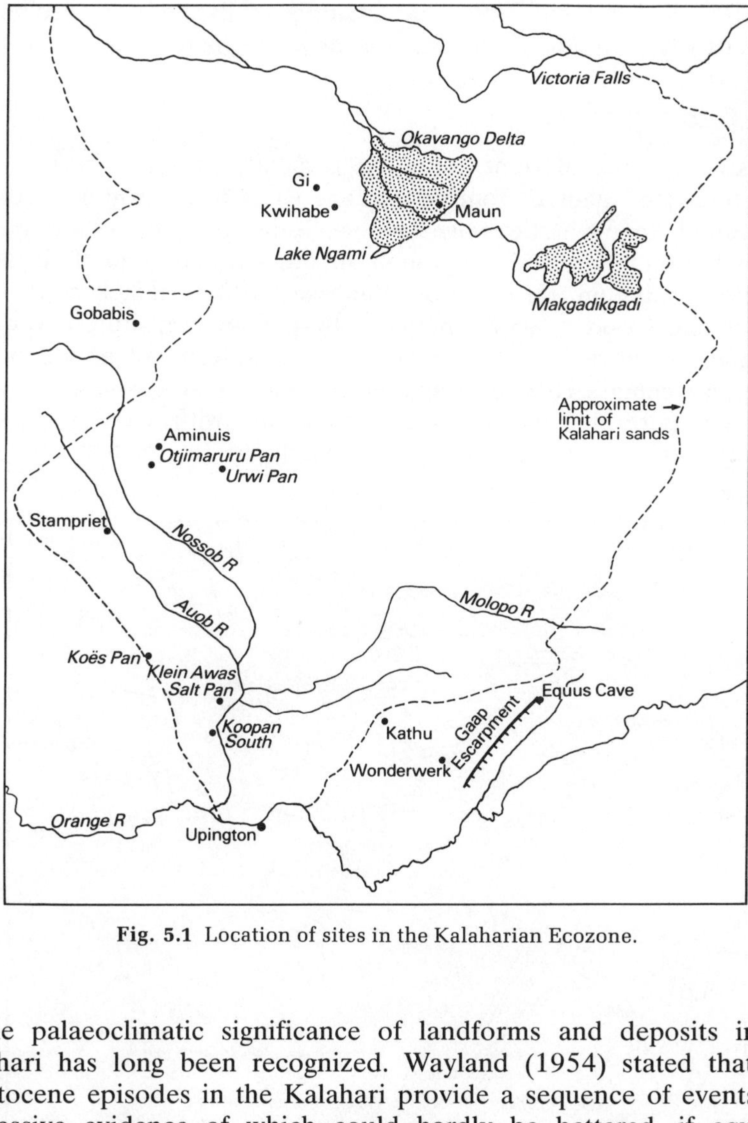

Fig. 5.1 Location of sites in the Kalaharian Ecozone.

The palaeoclimatic significance of landforms and deposits in the Kalahari has long been recognized. Wayland (1954) stated that 'the Pleistocene episodes in the Kalahari provide a sequence of events, the impressive evidence of which could hardly be bettered, if equalled anywhere in Africa'.

Subsequent investigations have qualified this statement somewhat and have established that, although the amplitude of Quaternary climatic fluctuations in the region has been considerable, they have ranged about a semi-arid mean (Lancaster 1984*b*). Generation of a true sequence of climatic changes has been hampered by the poor exposure and scattered nature of many of the Quaternary deposits. Correlation between areas has been made difficult by the uncertainties of dating many sequences, especially where radiocarbon dating of carbonate materials is concerned.

5.1.1 Dune systems

Extensive systems of dunes (Fig. 5.2), mostly of linear form, occur throughout the Kalahari from the Orange River at 20° south to latitude 12° south in Angola and western Zambia. Today, these dunes are fixed by savannah vegetation and are found in areas where rainfall is up to 800 mm per annum (Grove 1969; Lancaster 1981; Mallick *et al.* 1981; Thomas and Goudie 1984; Thomas 1984). They constitute one of the most impressive pieces of evidence for the widespread nature of late Quaternary climatic change in the Kalahari region.

The dunes form a massive semi-circular arc with a radius of some 1000 km, which corresponds approximately to the pattern of winds

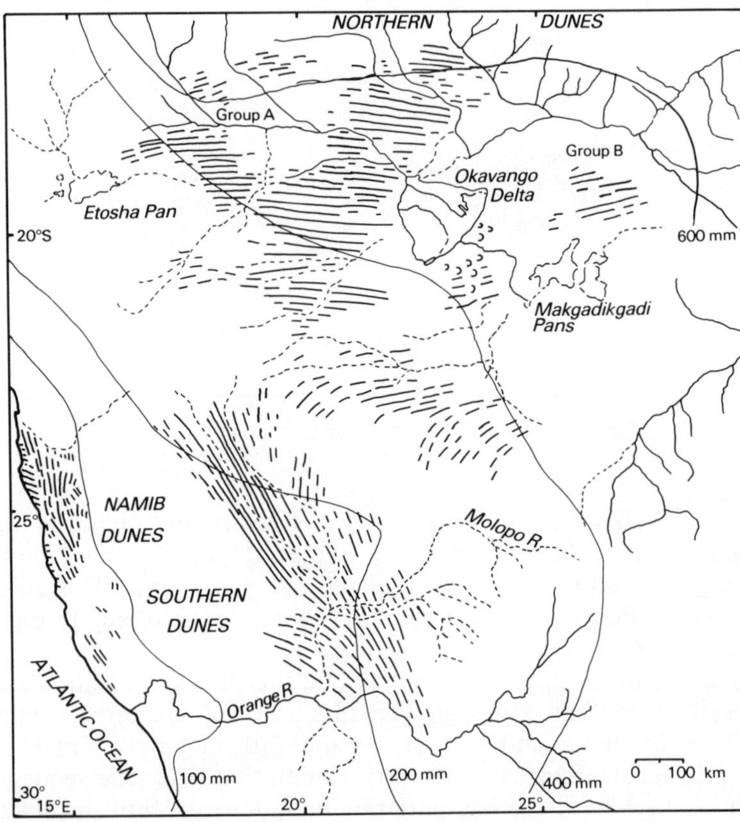

Fig. 5.2 Extent of fixed dunes in relation to isohyets (after Lancaster 1981).

outblowing around the Southern African anticyclone situated over the northern Transvaal. Within this arc, three distinct subsystems of dunes can be identified. The northern subsystem consists of dunes on east–west or east-north-east to west-south-west alignments. North and west of the Okavango Delta, they consist of broad straight parallel ridges up to 25 m high and 200 m long, with a spacing of 1.5–3 km. The dunes are very degraded and support a dense cover of tree savannah and open savannah woodland. In western Zimbabwe and adjacent areas of Botswana, the second subsystem consists of dunes with a relief of up to 25 m and 1.5–2 km apart (Flint and Bond 1968; Thomas and Goudie 1984). The southern dunes form the third subsystem and are best developed in a 100–200-km wide belt between the highlands of Namibia and the Orange River near Upington. They consist of north-north-west to south-south-east or west-north-west to east-south-east trending parallel to sub-parallel ridges 5–15 m high with a spacing of 0.2–0.4 km, with their steeper slopes facing south-west (Lewis 1936; Lancaster 1987). In many respects, these dunes can be considered almost identical with those in the Simpson Desert of Australia (Buckley 1981).

The palaeoclimatic implications of the fixed dune systems were first recognized by Flint and Bond (1968), and Grove (1969) and are discussed by Lancaster (1981, 1987), and Thomas and Goudie (1984). The existence of dunes in areas which today receive up to 800 mm of rainfall provides evidence for the former widespread extent of aridity in Southern Africa. Similar Late Quaternary fixed dune systems have been described from the Sahel region of Africa (Grove and Warren 1968; Talbot 1980, 1984), South America (Klammer 1982; Thomas and Goudie 1984), India (Goudie *et al.* 1973; Wasson *et al.* 1983) and Australia (Bowler and Wasson 1984; Jennings 1968). Many of these dunes were apparently last active during the Last Glacial Maximum period (Sarnthein 1978).

Lancaster (1981) compared the distribution of fixed dunes with the position of the 150 mm isohyet (Fig. 5.2), which he regarded as the limit of currently active dunes in Southern Africa and suggested that, when the northern subsystem of dunes was formed, the 150-mm isohyet must have lain 1000–1200 km north-east of its present position. This compares with shifts of 500 km reported from the southern Sahara (Grove and Warren 1968), 350 km in the Thar desert, and 900 km in Australia. In the case of the southern subsystem of dunes, the inferred changes are less at 200–250 km.

However, the mobility of dunes is not simply a function of low precipitation. It is inversely proportional to vegetation cover, but directly related to wind velocity. In turn, vegetation cover is a function of the ratio between potential evapotranspiration and rainfall. Various indices have been utilized to determine mobility thresholds for fixed dunes in the Sahel (Talbot 1984), Australia (Ash and Wasson 1983; Wasson 1984) and Southern Africa (Lancaster 1987), and from these to suggest the range of climatic conditions under which they would have been active. In the Kalahari, there is a gradient in sand mobility, as defined by the ratio between rainfall effectiveness and the amount of time the wind is able to move sand, from the northern and eastern parts of the south-western Kalahari Sand Sea to the currently active Namib Sand Sea. A threshold between clearly active and stabilized dunes in Southern Africa can be identified in a similar manner to the relationship derived by Wasson (1984) for Australian sand seas. Significant increases in wind velocity or changes in the moisture balance were necessary for dunes in the south-western Kalahari to be reactivated.

It is probable that the dunes in this part of the Kalahari were potentially active in the period 20 000–17 000 BP (Heine 1982; Lancaster 1987). At this time in Southern Africa, mean annual temperatures were some 5°C colder than at present (Heaton *et al.* 1986*a*). Newell *et al.* (1981) estimate that wind velocities at the latitude of the south-western Kalahari would have been 117 per cent of present values during this period. An increase in wind velocity of this magnitude would have substantially increased the sand moving power of winds in the region. Although evaporation would have been reduced as a result of lower temperatures, the increased windiness in the region would have been sufficient to reactivate dunes over most of the south-western Kalahari. Similar conclusions were reached by Wasson (1984) for Australian sand seas. However, to reactivate dunes in the northern and eastern parts of the south-western Kalahari, as well as those in the northern Kalahari, changes in the moisture balance would have also been necessary. A reduction in mean annual rainfall of at least 50 per cent would have been required to reactivate dunes throughout the region during this period. During the mid-Late Holocene period of aridity, temperatures in Southern Africa may have been 1–2°C warmer than at present (J. Deacon *et al.* 1984*b*), but wind velocities were similar to those of today. In these conditions, increases in the precipitation/evaporation ratio due to increased evaporation rates would not have been sufficient to reactivate dunes. Reactivation of dunes could only have taken place with a decrease in rainfall in the south-western Kalahari to 30–50 per cent of present values.

The changes in rainfall amounts indicated above imply major changes in circulation patterns over the Southern African subcontinent. Lancaster (1981) identified three patterns of palaeo-winds and sandflows in the Kalahari (Fig. 5.3) corresponding to each subsystem of dunes. Trends of the north-western group of dunes suggest that, when they were

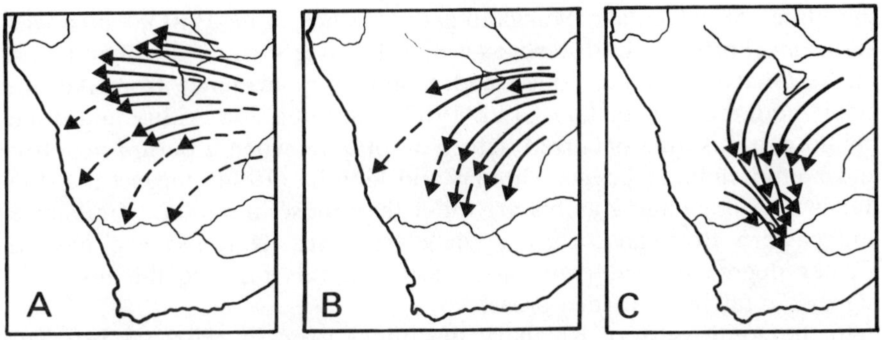

Fig. 5.3 Palaeosand flow patterns in the Kalahari. (A) Group A of northern dunes. (B) Group B of northern dunes. (C) Southern dunes (from Lancaster 1981).

formed, south-easterly and easterly winds penetrated much further into the interior of Southern Africa than they do today, implying a much stronger anticyclonic circulation, with a larger circulation radius than at present. The alignments of dunes of the second subsystem of dunes indicate that they were formed by easterly to north-easterly winds. These now occur much further to the south, in the northern Cape Province, suggesting that the year-round circulation resembled that of the modern July pattern and, in addition, lay 2° or 200 km north of its present position. The southern group of dunes have alignments which are the result of winds outblowing around both the Southern African and South Atlantic anticyclones. When these dunes were formed, winds were dominantly from the north and north-west, a situation which is reached in October today and in modern drought years (Tyson 1981). In addition, it is clear that periods of dune formation were also windier than at present. Particularly in the northern Kalahari, modern winds are too weak to move significant quantities of sand (Heine 1981; Thomas 1984).

Comparison of the patterns of modern and former dune forming winds suggests that periods of aridity and dune formation were associated with and were probably caused by increased persistence and strength of anticyclonic circulations, with annual circulation patterns closer to modern spring and winter situations. This, in itself, would probably have been sufficient to reduce rainfall by preventing the influx of moist air from the Indian Ocean, as occurs in modern drought years.

5.1.2 Chronology of dune building episodes

It is probable that the surface Kalahari sands have been redistributed by the wind at a number of times during the Quaternary (Grove 1969). The northern subsystems of dunes are truncated by the 920- and 945-m shorelines of palaeolake Makgadikgadi (Mallick *et al.* 1981), which was at this level prior to 18 000 BP, as well as by river systems north and west of the Okavango delta. Pollen and sedimentary evidence from Angola (Van Zinderen Bakker and Clark 1962) indicates that aridity and dune building in this area occurred before 38 000 BP, when a period of warm and humid climates began. Thomas and Goudie (1984) suggest that the dunes in Angola and Zambia are older than those in western Zimbabwe and eastern Botswana, because they are more degraded and have a greater degree of pedogenic alteration, but this may be the result of higher rainfall in the former areas.

In the south-western Kalahari, the dunes were in existence prior to 33 000–28 000 BP when they were stabilized by vegetation allowing the formation of organic soil horizons (Heine 1982) and calcrete pedotubules (Lancaster 1987). Heine (1981, 1982) and Lancaster (1978*b*, 1987) recognize periods of increased aeolian activity 20 000–17 000 BP, 12 000–9000 BP, and again in the mid-Holocene.

5.2 The northern Kalahari

5.2.1 The Makgadikgadi depression

Today, the Makgadikgadi depression (Fig. 5.1) contains a series of large pans which are ephemerally or seasonally flooded by inflow from the highlands of north-east Botswana or the Okavango delta via the Boteti River. The existence of shorelines indicating former high lake levels was first recognized by Grove (1969). Subsequently, shorelines at elevations of 945, 920, and 912 m have been identified and mapped by Cooke and his co-workers (Grey and Cooke 1977; Cooke 1980, 1984; Cooke and Verstappen 1984). The shorelines indicate the existence of very large lakes in the basin, with a maximum area of 37 000 km^2 at their highest level, although lacustrine deposits cover a total area of 60 000 km^2 in the Makgadikgadi and Okavango basins (Cooke 1984). Geomorphic evidence discussed by Helgren (1984) indicates that the 920 and 912 m lakes were in existence for considerable periods of time.

H. J. Cooke (1984) reports radiocarbon dates from a variety of calcretes and lacustrine carbonates associated with lake shorelines. Although the interpretation of the dates is often difficult, Cooke (1984) argues that Makgadikgadi depression held lakes at the 945 m level between 40 000 and 35 000 BP, and at the 920 m level on a number of occasions between 26 000 and 10 000 BP. However, Helgren (1984) notes the association between Earlier Stone Age artefacts and the 945 m lake level, suggesting that it is probably of mid-Pleistocene age. Heine (1978) reported a series of eight radiocarbon dates from lacustrine sediments and molluscs in the Makgadikgadi area which cluster between 30 250±520 BP (Hv-8352) and 19 170±660 BP (Hv-8366). It is difficult to assign these dates to specific shoreline elevations, but correlation with the 920 m level seems probable. According to Heine (1978), the lake was again at a high level, possibly at 912 m, around 11 920 BP (Hv-8367). Low lake levels and dry conditions are inferred by Cooke (1984) for the period between 35 000 and 26 000 BP and around 19 400 BP when peat formation in small pans indicates that it was also cooler than at present. The lake level reached 912 m or 2 m above its present elevation around 3130 BP, when it was fed by a strongly flowing Boteti River, and again between 1700 and 1600 BP. In the south-western parts of the Makgadikgadi basin, Later Stone Age settlement was associated with shallow freshwater lakes near the mouth of the Boteti River prior to 2960 BP (Helgren 1984).

Ebert and Hitchcock (1978) have calculated that rainfall increases of two to three times present levels would have been necessary to sustain lakes at the 920- and 945-m levels, given present temperature conditions. Cooke (1980) argues for a rainfall of 1500 mm on the lake itself, together with substantial increases in inflow, ultimately derived from Angola, via the Okavango system. Caution has been urged by Grove (1969) and Cooke (1980) in the palaeoclimatic interpretation of high lake levels in the Makgadikgadi. The front of the Okavango delta is known to be tectonically active and a small change in elevations in this low relief area could cause radical drainage diversions. Grove (1969) has suggested that only the Zambezi River has a discharge large enough to maintain a lake at the 945-m level and Helgren (1984) speculates that the 945-m lake may have been a short lived catastrophic event, resulting from a unique combination of tectonic and climatic events in the mid-Pleistocene.

5.2.2 The Okavango region

The Okavango delta is a complex hydrological system responding to slow tectonism along the Thamalakane fault which defines its south-eastern margins, and to changes in rainfall both locally and over its Angolan catchment area. Evidence for past hydrological changes has been identified by Grove (1969) and Cooke (1980). Shaw (1986*a*) has identified lake shorelines at elevations of 936 m in both the Ngami and Mababe basins, which lie at elevations of 919 m at opposite ends of the Thamalakane fault zone. At this elevation, the Okavango would have supported a 5000 km^2 lake, with an outflow to the Zambezi River via the Savuti area, as well as down the Boteti to the Makgadikgadi. A local rainfall of 200 per cent of the present amount would have been sufficient to maintain such a lake level (Shaw 1986*a*). A further shoreline at 932 m marks a trend to closed lakes in these basins. Radiocarbon dates from lacustrine carbonates reported by Shaw (1986*a*) indicate that a lake was at the 936-m elevation between 17 000 and 13 000 BP. Lake levels were low in the early Holocene, but reached 932 m around 6000 BP. A return to the 936-m level in the Late Holocene is also suggested.

5.2.3 Gi and Xangwa

Deposits at Gi and Xangwa in western Ngamiland preserve a long history of deposition in the Dobe valley in climates both wetter and drier than at present, which is set against a background of slow small scale tectonism associated with the development of the Okavango graben (Helgren and Brooks 1983).

The valley was established in the Early Pleistocene across a landscape of large linear dunes. Uncemented alluvial and colluvial deposits containing MSA occupation debris indicate a semi-arid climate in the area during the early Late Pleistocene, with an ephemeral pond in the valley. Pollen analyses indicate that vegetation at this period was an open treeless grassland, with swampy conditions in the valley (Kuman 1985). Middle Stone Age settlement retreated to the valley sides as a lake up to 9 m deep rapidly developed. The lake is marked by a thick lacustrine limestone. With coeval spring and channel tufas in the Xangwa valley, it represents a long continued humid period prior to 31 000 BP. Desiccation of the lake occurred in early Later Stone Age times. The valleys were occupied by small pans, with small dunes on interfluves, indicating a climate more arid than today. A further lacustrine limestone was deposited in a 4-m deep lake 23 000–22 000 BP, indicating a return to wetter conditions. As with the earlier humid period, tufas also formed in the Xangwa valley, indicating increased discharge of karstic waters from the Aha hills. The period spanning the Last Glacial Maximum appears to have been drier than at present, with extensive colluvia, which were later calcreted, surrounding small pans in the valley. Holocene sands, silty and sandy clays, and clays become increasingly humic in later periods, suggesting an increase in sedimentation rates and the production of organic matter brought about by increased local rainfall. In the Late Holocene, drier episodes resulted in renewed deflation to form the modern pan at Gi.

5.2.4 The Kwihabe Hills

The record of palaeoclimates contained in deposits and landforms in the Kwihabe Hills, located 50 km south of Gi and Xangwa, contains further evidence for alternating wet and dry climates in this area (Cooke 1975, 1984).

The valley of the Kwihabidum flows eastwards across the linear dunes and bisects the Kwihabe Hills in a 10–15-m deep gorge-like section. Cooke (1980, 1984) identified four distinct calcified aeolian sand and fluvial gravel units in the valley. The first two units have radiocarbon dates of greater than 45 000 BP, but the later units have consistent dates between 11 000 and 10 000 BP.

Drotsky's Cave lies in the largest of the Kwihabe Hills. Its linear development clearly indicates a phreatic origin related to former groundwater levels before the local incision of the Kwihabidum valley. Extensive speleothems occur in the cave and can be divided into four major units. Radiocarbon dates on sinter II range between 34 000±1700 BP (Wits-306) and 29 300±1000 BP (Wits-377), suggesting to Cooke (1984) that humid conditions prevailed in the area prior to 30 000 BP. A series of eight dates on sinter III clusters between 16 000±200 BP and 13 000±200 BP (Wits-164 to 169). During this period, stalagmite growth was 0.4–1.0 mm per year, indicating very humid conditions and a rainfall three times that of the present (Grey and Cooke 1977). Further speleothem development occurred in wetter climates between 2500±100 and 2000±100 BP (Wits-122, 123), and again around 750 BP (Wits-195). Between deposition of sinters II and III, invasion of the cave by aeolian sands indicates a period of dry climates (Grey and Cooke 1977).

5.2.5 Etosha Pan

Etosha Pan represents a large palaeolake basin in the north-western Kalahari in Namibia, created by the fusion of smaller pans by scarp retreat processes (Rust 1984). Stratigraphic studies, backed by up to 35 radiocarbon dates enabled Rust (1984) to put forward a palaeoenvironmental interpretation of its Late Pleistocene and Holocene development. Radiocarbon dates on pedogenic calcretes cluster between 33 000 to 27 000 BP, 22 000 to 18 000 BP, and 10 000 to 9000 BP suggesting periods of geomorphic stability at these times. Evidence for increased precipitation and higher water levels in the pan is provided by lacustrine carbonates with an age of 11 900±120 BP (Pta-3401) and spring deposits at Namutoni, dated to 9310 ± 90 BP (Pta-3043).

5.3 The southern and south-western Kalahari

The southern and south-western Kalahari is the driest part of the region, with a mean annual rainfall of 150–300 mm. Its landforms display a complex variety of fluvial, pan and aeolian affinities, suggesting that the region has reacted sensitively to climatic changes. Major river valleys (the Auob, Nossob) enter the south-western Kalahari from the highlands of Namibia and south-western margins of the Kalahari (the Molopo and the Kuruman). Flow today is ephemeral or seasonal in the upper reaches of these streams and floods reach the end point of the Molopo system only after periods of exceptional rainfall (e.g. 1933, 1974).

5.3.1 Pans

This region contains some of the largest concentrations of pans in Southern Africa. The first group of pans lies along the watershed between the Molopo and Nossob drainages, and the fossil river valleys which lead north to the Makgadikgadi Depression (Lancaster 1978*b*). The second concentration of pans occurs in a belt which extends from west of Koes in Namibia to north-west of Upington on the Orange River. Within this belt, large pans are particularly numerous in the vicinity of the lower Molopo valley [the Mier country of Du Toit (1926)]. Pans in both areas provide a partial record of Late Quaternary climatic change in the region.

On the southern margins of most pans, except those south-east of the Molopo valley, lie areas of lunette dunes, which rise to elevations of as much as 50 m above the adjacent pan floors (Fig. 5.4). In the vicinity of the lower Molopo valley, linear dunes also extend down wind from the lunettes (Fig. 5.4). In many areas, there are two main lunettes: an outer dune of red brown sand, and an inner dune of grey brown, slightly calcareous sand and clayey sand (Lancaster 1978*b*). At many pans in the south-west of the region there is a third lunette close to the pan edge which is composed of sandy clay (Lancaster 1986). The lunettes represent episodes of deflation of pan sediments. In the southern Kalahari, Lancaster (1978*b*) suggested that the outer dune was composed of the surface Kalahari sands deflated to form the original pan depression, whilst the inner dune represented deflation from the pan floor, following an intervening period when small lakes occurred in the pan depressions.

The conditions under which lunette dunes may form have been summarized by Bowler and Wasson (1984). Strong winds are required to deflate material from pan surfaces, but high water tables or frequent flooding followed by high evaporation are also needed to break up surfaces by salt weathering and to mobilize pelletal clays for clay dune formation. Many pans, especially in the south-western Kalahari, are located in areas where groundwater is close to the surface and Lancaster (1986) has suggested that periods of lunette formation may have been associated with higher local water tables. The alignments of the lunettes

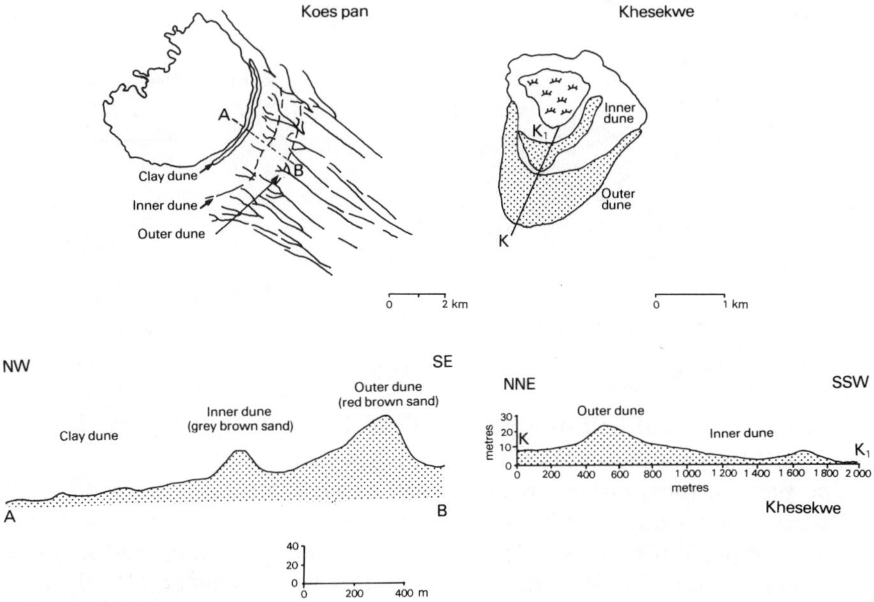

Fig. 5.4 Characteristics of Koes and Khesekwe pans and their fringing dunes (after Lancaster 1978*a*, 1986).

show that in all areas both sandy lunettes were formed by winds from similar northerly or west-north-westerly directions, but formation of clay dunes was by north-westerly winds. Present-day dry season winds in the area are north-east or north, swinging around to north-west in late winter and spring. A much more extreme and fluctuating climate with higher water tables than occur today seems to be implied by the formation of lunettes in the southern and south-western Kalahari. The evidence from the alignments of the lunettes tends to confirm that from dune patterns which indicate that periods of increased aeolian activity were associated with a greater frequency of northerly and westerly winds throughout the region.

Dating of the periods of lunette formation is difficult. The innermost clay lunette at Koopan Suid had ceased formation by 1000 BP, as indicated by a radiocarbon date of 960±140 BP (Pta-3865) on a hearth situated on the pan-ward slopes of this dune. The main period of clay dune formation probably took place during the Late Holocene, when there is evidence for warmer and drier climates in the region (J. Deacon *et al.* 1984*b*). Formation of the inner lunettes may have taken place following the period of high water levels in pans which occurred between 16 000 and 12 000 BP, on the basis of stratigraphic evidence from Urwi Pan (Lancaster 1979) and elsewhere (Fig. 5.5). The outer lunette may

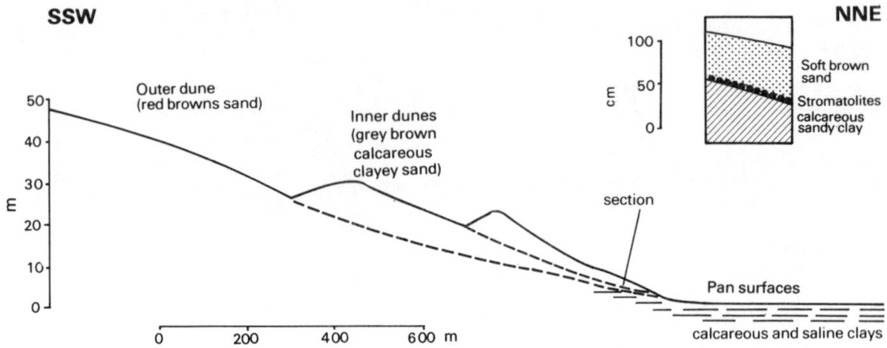

Fig. 5.5 Stratigraphy and occurrence of stromatolites at Urwi Pan (after Lancaster 1979).

have formed coeval with the period of aeolian activity between 20 000 and 16 000 BP, but there is no direct evidence to support this.

The pans in the southern Kalahari are underlain by 2–3 m or more of calcareous and saline sandy clays which cover areas two to three times that of the present pans. Evidence for permanent wave-agitated lakes at Urwi Pan around 16 000 BP was presented by Lancaster (1979) who suggested that rainfall in the southern Kalahari was 1.5–2 times present amounts during this period. Fifty kilometres west of Urwi Pan, relict spring deposits on the margins of pans indicate increased groundwater discharge and higher groundwater levels. At Aminuis, spring tufas have radiocarbon dates of 33 500 and 21 900 BP suggesting a fall in groundwater levels of 2 m during this period. Nearby, at Otjimaruru Pan, algal mats and stromatolitic spring tufas have radiocarbon dates of 30 400 and 23 700 BP, and calcified organic sediments are dated to 12 050 BP. Butzer (1984*a*) reports dates of 17 200 and 11 800 BP on spring deposits from pans in the Gobabis area. Taken together, this evidence suggests that water levels and groundwater discharge were high around 33 000–30 000 BP, and again between 17 000 and 11 000 BP in this part of the Kalahari.

Studies of the isotopic composition of groundwater in the Stampriet aquifer by Heaton *et al.* (1983) show that the concentration of nitrates and 'excess air' in the groundwater has varied through time, reflecting parallel environmental changes. The concentration of 'excess air' reached a peak between 30 000 and 26 000 BP, and again between 14 000 and 8000 BP when the concentration of nitrates also reached a peak. On the basis of these observations, Heaton *et al.* (1983) suggest that periods of increased rainfall occurred in the western Kalahari between 35 000 and 26 000 or 20 000 BP, and again between 14 000 and 8000 BP.

In the south-western part of the Kalahari, pan sediments are rarely more than 1 m thick and overlie bedrock. This suggests that either deflation has been more effective in this region or that the period of higher water levels did not occur there. Heine (1982), however, reports a date of 23 000 BP on stromatolite concretions from Klein Awas Pan. In the lower Molopo valley, there is sedimentary evidence to suggest that flows were perennial or at least seasonal prior to 19 000 BP (Heine 1981, 1982). There was a further period of increased flow and seasonally wet conditions between 16 000 and 13 000 BP.

5.4 The south-eastern Kalahari

This area is transitional between the semi-arid Kalahari and the more humid highveld regions to the east. Sites in this region are therefore sensitive indicators of shifts in the boundary between these ecozones.

5.4.1 Kathu Pan

Studies by Beaumont *et al.* (1984) of the sediments infilling dolines and palaeochannels developed in 50-m thick calcretes at Kathu Pan in the northern Cape provide an interesting record of Late Pleistocene climatic change in the region. The pan today is covered by short grass with thickets of *Acacia karroo* which merge to the south with open *Acacia erioloba* woodland and more open grassland to the west.

The calcrete which forms the substrate of the pan is mantled by 1.8–2.3 m of unstratified Holocene peats and sands, whilst the dolines and palaeochannels contain up to 4–6 m of silty sands and peats, the lower members of which contain Earlier and Middle Stone Age artefacts. Sedimentary and palynological data are consistent and indicate a sequence of environmental changes which reflect long-term groundwater trends (Butzer 1984*a,b*). The chronology of the upper units is controlled by numerous radiocarbon dates (Fig. 5.6).

A stable period of relatively dry conditions, with a slow rate of sediment accumulation and a water table 2 m lower than the present prevailed from before 32 000 BP to around 11 000 BP. The presence of well sorted fine sands in the deposits indicates a strong primary or derived aeolian component (Butzer 1984*a,b*). Between 11 000 and 7400 BP, a marked increase in calcrete clasts and a reduced organic content of the peaty layers suggest increased sheet erosion from the surrounding area, as a result of a disrupted vegetation cover and markedly drier conditions. Pollen spectra from the period 7400–4500 BP (Zone I, with a basal date of 7350±90 BP, Pta-3073) indicate an open grassland with locally wet habitats. There is evidence of a more varied tree cover, suggesting that the climate was both warmer and wetter than the present, with a high water table permitting peat accumulation on the pan surface. From 4500 to 2700 BP (Zone II), the climate was drier than the present, with brackish water and seasonal mud flats in the pan. There is a virtual absence of arboreal pollen and species indicating wet habitats. A grassland with locally wet conditions has existed in the area throughout the last 2700 years (Zone III).

5.4.2 Wonderwerk Cave

This 140-m deep dolomite cave, situated on the eastern margins of the Kuruman Hills, contains a complex series of deposits with a cumulative thickness exceeding 5 m, with a basal Middle Pleistocene age (Beaumont *et al.* 1984).

Sedimentary evidence (Butzer 1984*a,b*) indicates that the period prior to 30 000 BP was generally dry. Around 35 000 BP there was a catastrophic roof collapse. Sediments of the following periods indicate increased frost action, probably in very cold climates. A major period of frost action occurred between 35 000 and 26 000 BP. Weathering of the frost derived clasts, indicating greater humidity, increases after 25 000 BP. The absence of deposits spanning the Last Glacial Maximum period led Butzer (1984*a,b*) to argue for a period of erosion by sporadic heavy rainfall in a climate considerably wetter than today. This interpretation is at variance with indicators of a dry climate at nearby Kathu Pan.

Deposits dating to later than 13 500 BP also contain pollen and micromammal remains, which were studied by Van Zinderen Bakker (1982*b*) and Avery (1981), respectively. Van Zinderen Bakker (1982*b*) identified three pollen zones: I (9200–5000 BP) with spectra characterized by a marked increase in Compositae between 7430±60 BP (Pta-2798) and 5870±70 BP (Pta-2545) indicating a climate wetter and warmer than today and open savannah vegetation of tropical affinities; Zone II dates from 5000 to 3–2000 BP, with the pollen indicating an almost treeless grassland and dry climates; Zone III dates between 2000 and 1000 BP, and the inferred vegetation suggests a climate wetter than Zone II and similar to today.

Avery (1981) suggests a different sequence of climates, based upon the community structure of the micromammals. The period 10 000–9000 BP was characterized by dry open scrub and few trees and bushes. Around 6000 BP, conditions were moister with short grass, and more trees and bushes. Maximum precipitation effectiveness was reached at 5000 BP when savannah woodland was present in the area. The period 5000–4000 BP was drier, although with a slight increase in effective precipitation at 4000 BP, marked by an increase in tree and bush cover. Studies of large mammals (Thackeray 1984) indicate an increase in primary production between 10 000 and 3000 BP, with fluctuations thereafter.

Beaumont *et al.* (1984) have put forward a reinterpretation of the Late Pleistocene and Holocene pollen, and micromammalian data, based upon a revised temporal framework provided by more than 30 radiocarbon dates (Fig. 5.6). From 13 000 to 11 500 BP, the abundance of frost weathered debris indicates a climate markedly colder than at present. Between 10 500 and 9000 BP, open scrub savannah vegetation was present, with trees along drainage lines. Drier conditions are inferred, with more regular and intense frosts indicated by an increase in roof spall relative to the present. The period 8000 to 6000 BP was one of higher and more effective precipitation with more short grass and an increase in the abundance of trees and shrubs, such that a savannah woodland existed at the peak of warmth and humidity between 6000 and 5000 BP. Enhanced cave humidity is indicated by a high fern spore count and a maximum expansion of the stalagmite boss. After 5000 BP,

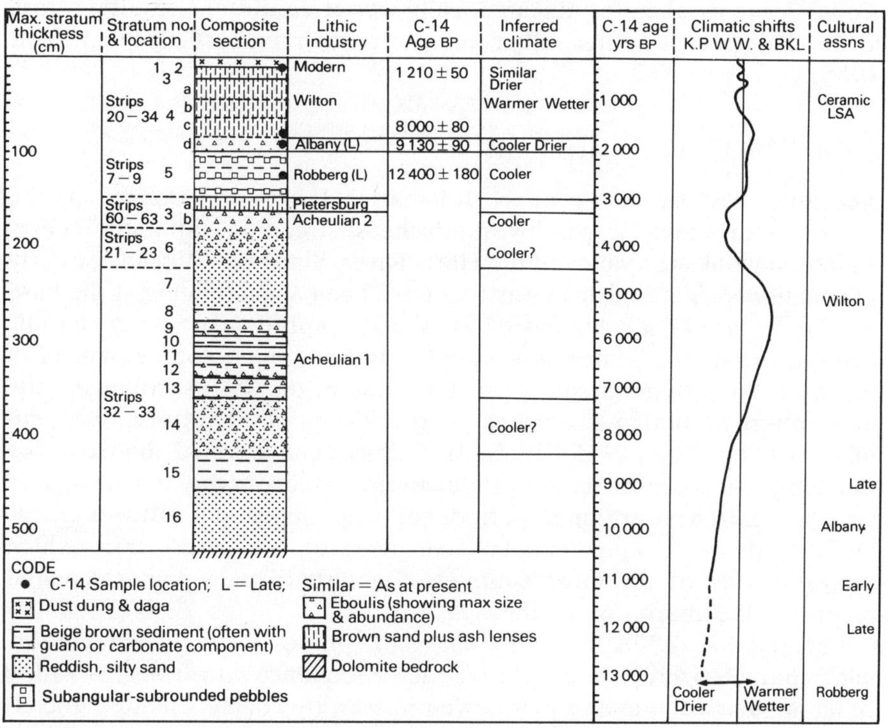

Fig. 5.6 Wonderwerk Cave composite stratigraphy (left) and comparison between inferred climatic shifts over the last 13 000 years at Kathu Pan (KP), Wonderwerk (WW), and Blinkklipkop (BKL) (right) (after Beaumont *et al.* 1984). (Reprinted from: *Late Cainozoic palaeoclimates of the Southern Hemisphere*, (ed. J. C. Vogel) 1984. A. A. Balkema, P.O. Box 1675, Rotterdam, Netherlands.)

climates became drier and cooler. Savannahs were replaced by treeless grassland, although woody vegetation increased slightly after 4000 BP. From 2500 to 1000 BP the pollen data indicate that climates were at times wetter than today, with travertine deposition at *c.* 1950–1450 and *c.* 850–750 BP.

5.4.3 Equus Cave

Pollen in Holocene strata from this locality on the Gaap Escarpment studied by Scott (Beaumont *et al.* 1984) indicate a Kalahari-type thornveld similar to that which exists today in the period between 7480 ± 80 BP (Pta-2495) and 2390±55 BP (Pta-2452). The underlying Late Pleistocene strata have pollen suggesting fewer trees and the presence of small shrubs, indicative of a cooler climate than that after 7500 BP. This inference is supported by the greater than present mean size of *Canis mesomelas* and *Vulpes chama* (Klein 1986). The presence of the aquatic *Kobus leche* is taken to indicate locally wetter conditions resulting from lowered evaporation rates rather than increased rainfall (Grine and Klein 1985).

5.4.4 The Gaap Escarpment

The south-eastern margin of the Kalaharian Ecozone is defined by the Gaap Escarpment, a southwest–northeast trending dolomite cuesta which overlooks the valley of the Harts River. Since the Miocene or early Pliocene, deposits of calcareous tufa have been accumulating at the base of the escarpment, deposited as sheets (aprons) or by waterfalls (carapaces) at springs or below streams draining the dolomite plateau to the west. Attention was drawn to the tufas in the 1920s, following the description of hominid remains of the Taung child (*Australopithecus africanus*) by Dart (1925). Peabody (1954) demonstrated the existence of a long and complex record of Pleistocene events in the tufa sequences and recognized four main periods of tufa deposition. Butzer *et al.* (1978*b*) describe four regional sequences of tufa deposition, which represent one of the most complete Quaternary palaeoenvironmental records in the interior of Southern Africa.

Butzer *et al.* (1978*b*) identify six major phases of tufa accumulation, each characterized by a distinctive facies sequence which begins with a basal angular rubble, and is followed in turn by coarse rounded gravels and gritty calcreted sands, subhorizontally bedded wavy to laminated tufas, increasingly steeply bedded forest tufa carapaces, and terminated by surface and fissure corrosion, with an infill of sandy wash. This sequence is interpreted as being related to changes in precipitation amounts, intensity and duration (Butzer *et al.* 1978*b*; Butzer 1984*a,b*). Lower temperatures and evaporation rates would also facilitate tufa deposition. The sequence of events begins with more frequent high intensity rains which erode soils and surface sands, and deposit them as colluvial aprons or fillings in karst fissures. It is followed by intense frost weathering of dolomites to produce angular debris at the base of cliffs which today lack such deposits. The angular debris is then reworked by repeated high energy run-off. As time goes on, aquifers are recharged so that stream flow becomes more persistent. Less debris is transported and sheets of calcium carbonate begin to accumulate on footslopes. When discharge becomes nearly perennial, waterfall tufas build out over earlier tufa sheets, with corrosion in periods of maximum flow. The end of the cycle is marked by decreased rainfall which leads to the disintegration of the drainage network and restriction of tufa accumulation to spring seeps, with corrosion of joint systems.

Tufas I and II are assigned a Tertiary age and are tentatively correlated with the Wedberg terraces of the Older Gravels of the Vaal River gravels (Helgren 1978*a,b*). Tufa III is regarded as late Early Pleistocene and Tufa IV probably dates to the Middle Pleistocene. Blue Pool Tufa I at Taung was deposited in the period 103 000 to 30 000 BP as indicated by bracketing U/Th ages (Vogel and Partridge 1984). Subsequently, but prior to depositon of Blue Pool Tufa II in a period of protracted spring activity 33 000–14 000 BP, a deep gorge was re-excavated, a deep red palaeosol formed on the Gaap Plateau and rockfalls accumulated below the escarpment (Butzer 1984*a,b*). Deposition of Blue Pool Tufa II was preceded and followed by significant mechanical weathering to produce angular rubbles suggesting to Butzer (1984*b*) that increased frost action occurred at this time. It is not clear if tufa carapaces at Gorrokop and Mazelsfontein, with dates ranging between 26 840 and 22 000 BP, represent an earlier sub-stage of this phase of tufa deposition or are substantially older (Butzer 1984*b*). Phase IV includes three episodes of Holocene tufa deposition and increased spring activity with radiocarbon dates between 9600–6500, 3200–2400, and 1300–250 BP. Pollen from archaeological deposits indicates, however, that there was little change in the vegetation (Butzer 1984*b*).

Today, rainfall in the vicinity of the Gaap Escarpment averages 300–400 mm annually and tufa deposition takes place only in years of exceptional rainfall. Significant tufa formation appears to take place only when rainfall is in the range 600–1000 mm per year. In the eastern Transvaal, Marker (1972) reports active tufa formation in areas where rainfall is 600–750 mm per year. Thus, tufa formation on the Gaap Escarpment is interpreted by Butzer *et al.* (1978*b*) as indicating conditions of increased rainfall and a sub-humid climate. The volume of

Late Pleistocene tufas decreases southwards, parallel with the modern decrease in rainfall totals, indicating to Butzer (1984*b*) that the seasonal distribution of rainfall in periods of tufa formation was similar to the present.

5.5 Chronological summary for the Kalaharian Ecozone

Despite the size of the Kalaharian Ecozone and the consequent diversity of environments, consistent palaeoclimatic trends are evident for much of the Late Pleistocene.

There is little precise information on palaeoclimates in this region during the early Late Pleistocene (129 000–50 000 BP), but analysis of the lowermost deposits at Wonderwerk and Kathu may yield some useful data.

During the period 50 000–24 000 BP the dating of deposits and landforms is problematic. There are indications from Gi of dry climates in the earlier part of this period. Latterly, high lake levels in the Makgadikgadi and speleothems from Kwihabe indicate wetter conditions, but dating control is poor. The south-eastern Kalahari appears to have been dry and increasingly cold, with increased frost action on the Gaap Escarpment.

During the period 32 000–22 000 BP there is evidence for widespread humid conditions in the Kalahari with intervals during which lake levels were lower. In the south-western Kalahari, dunes were stabilized by vegetation, allowing soils to form, and groundwater levels were high. In the Okavango and Makgadikgadi, lake levels were high with a drier interval *c.* 25 000 BP in the Okavango and *c.* 20 000 BP in the Makgadikgadi (Shaw 1986*b*). It was cold and moist in the south-eastern parts of the region, with increased spring discharge on the Gaap Escarpment leading to tufa formation. At the end of this period, lake levels in the northern Kalahari and elsewhere fell rapidly.

During the period spanning the Last Glacial Maximum, it appears that climates throughout the region were dry and cold. However, tufa deposition in a subhumid climate continued on the Gaap Escarpment.

Between 17 000 and 11 000 BP there is evidence for increased moisture availability in both the north-west and the southern Kalahari, with high lake and pan levels, and increased stream flow and groundwater discharge. Conditions do not seem to have been as wet as those before 20 000 BP and lake levels in the Okavango and Makgadikgadi were lower than between 32 000 and 20 000 BP but higher than at present until around 12 000 BP. The south-east of the region remained cold and dry, although tufa deposition continued at Taung prior to 14 000 BP.

During the Holocene, climates in the region were generally warmer than in preceding millennia with fluctuating wet and dry phases. The terminal Pleistocene/early Holocene (12 000–8000 BP) appears to have been dry in the south-western and south-eastern Kalahari, and by 9500 BP conditions essentially similar to the present were established in the northern Kalahari. The mid-Holocene (8000–4000 BP) was a period of increased humidity and maximum warmth throughout the region. After *c.* 6000 BP the levels of Lake Ngami rose to overflow level (Shaw 1986*b*). Drier and cooler conditions characterize the late Holocene (4000–0 BP) with wetter intervals between 3000 and 2000 and between 1600 and 1000 BP in the southern part of the ecozone (Beaumont *et al.* 1984; Butzer 1984*b*) and from *c.* 4200–3600 and *c.* 2500–2000 BP at Drotsky's Cave, Lake Ngami and Makgadikgadi in northern Botswana (Shaw 1986*b*).

5.6 The Transvaalian Ecozone

The Transvaalian Ecozone (Fig. 5.7) is characterized by semi-arid acacia steppe and bushveld. For the most part it occupies the inland plateau of Southern Africa at altitudes of 1200–1800 m inland, and lower in coastal Natal and the eastern Cape. Annual rainfall is between 500 and 300 mm, and falls mainly in the summer. It comprises the area from northern Namibia in the west to the Natal coast in the east, and from southern Zimbabwe and north-eastern Botswana in the north to the eastern Cape coast in the south, but excludes the Drakensberg, inland Natal, and the Orange Free State. Because of its wide geographic range, the vegetation, climate, and zoogeography show some variability. It is driest in the west where ungulates recorded in historic times includes grazers such as blue wildebeest, Cape hartebeest, tsessebe, Cape buffalo, springbok, gemsbok, Burchell's zebra and warthog, browsers such as kudu, bushbuck, giraffe and black rhinoceros, and mixed feeders such as eland and impala. To the east, the bushveld is better watered with browsers and mixed feeders more common than grazers, and springbok and gemsbok absent. The narrow strip along the coast of Natal, Transkei and the eastern Cape east of the Great Fish River is sub-humid with tropical thornbush grading southwards to scrub forest. Many of the bushveld species amongst the ungulates, in particular the grazers, did not extend southwards in this part of the ecozone (Klein 1980).

The most complete Late Quaternary sequences are those from archaeological cave deposits. Even these preserve a patchy record, but they are useful in that they include not only sedimentological information, but also biological data.

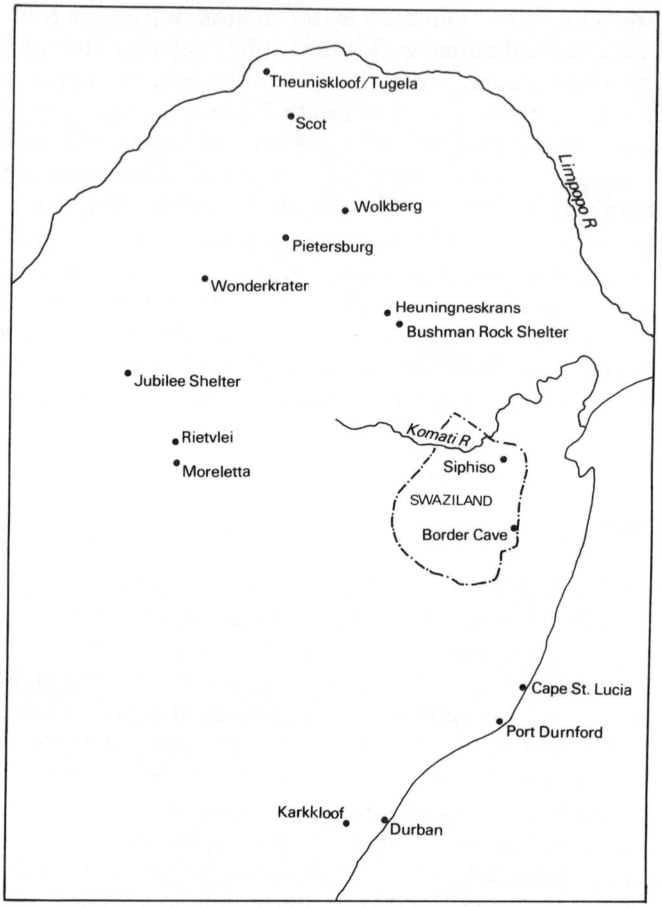

Fig. 5.7 Location of sites in the Transvaalian Ecozone.

5.6.1 Border Cave

Border Cave lies at an altitude of 600 m in the Lebombo Mountains on the border between Swaziland and KwaZulu. It contains a long sequence of Middle Stone Age assemblages and associated faunal remains, including early *Homo sapiens sapiens*. There is a long period of non-occupation between about 35 000 BP and Iron Age occupation dating within the last 1000 years. Detailed stratigraphic and sediment analyses permit the identification of eight Pleistocene sedimentary cycles which could span the last *c.* 200 000 years (Butzer *et al.* 1978*a*; Table 5.1).

Radiocarbon dates back to 49 000 BP have been used to determine sedimentation rates and although there are no oxygen isotope analyses from the site, units have been linked to the oxygen isotope sequences by inference. These extrapolated dates, with allowance for compaction, are the basis for suggesting that the lowermost deposits date to between 186 000 and 208 000 BP. Periods of greater cold are inferred from increased percentages of grit and angular rubble in cave sediments that are held to indicate increased mechanical weathering. Angular rubble is interpreted as representing periods of frost weathering. Butzer (1984*a*) suggests that it was relatively warm and wet *c.* 108–105 000 BP, with cold conditions during the Howiesons Poort phase of the Middle Stone Age *c.* 80 000 BP. This was followed by somewhat warmer or drier conditions and then éboulis sec indicating cold conditions *c.* 69 000 BP. Cold and dry conditions followed. It was wetter again *c.* 57–49 000 BP with frost action between 38 000 and 33 000 BP, and it was cool to cold and dry between *c.* 29 000 and 13 000 BP (Table 5.1).

Avery (1982*a*) has studied the micromammalian remains from Border

Table 5.1. Border Cave palaeoclimatic inferences based on sediment analyses (after Butzer *et al.* 1978*a*).

Level	Approximate age (years BP)	Interpretation
Level 1a	700 BP to present	Limited grit production. Modern conditions of sedimentation and climate. Iron age occupation.
Hiatus	*c.* 13 000–700	
Level 1b	*c.* 29 000–13 000	Moderate éboulis and grit production. Moderate frost weathering. Climate cool to cold, possibly dry.
Hiatus	*c.* 33 000–29 000	
Level 2–3	*c.* 38 000–33 000	Important éboulis horizon. Effective frost weathering. Climate cold, possibly wet. Early Later Stone Age occupation.
Hiatus	*c.* <49 000–39 000	
Level 4		Limited grit production. Climate similar to present, possibly wetter. Late Middle Stone Age occupation.
Hiatus	*c.* 57 000	
Level 5a		Moderate grit production. Climate cooler and very dry.
Level 5b		Major éboulis horizon. Effective frost weathering. Climate cold, possibly wetter. Late Middle Stone Age occupation.
Hiatus	*c.* 69 000	
Level 6–7a		Minimal grit production. Climate warmer or drier. Unoccupied at first, then Middle Stone Age.
Level 7b–8		Major éboulis horizon. Maximum effectiveness of frost action, becoming less with time. Climate cold, initially wetter. Howiesons Poort occupation.

Table 5.1. *(continued)*

Level	Approximate age (years BP)	Interpretation
Hiatus	*c.* 108–105 000	Minor soil horizon. Chemical weathering with gypsum mobilization. Greater cave moisture. Climate warm and wet.
Level 9		Eboulis and grit horizon, grading up into finer, laminated beds. Climate initially cold, becoming more temperate and/or drier. Early Middle Stone Age.
Level 10		Major éboulis and grit horizon. Effective frost weathering. Climate cold and probably wetter. Early Middle Stone Age.
Level 11		Moderate éboulis production. Frost weathering. Climate colder. Early Middle Stone Age.
Hiatus	*c.* 154–145 000	Major soil horizon. Protracted chemical weathering. Greater moisture. Climate warm and very wet.
Level 12		Moderate grit and éboulis production. Effective frost weathering. Climate cold. Early Middle Stone Age.
Hiatus	*c.* 165 000–154 000	
Level 13		Some éboulis production. Some frost weathering. Climate cool. Early Middle Stone Age.
Level 14–15		Some grit production at first, but sediments comparable to 1a. Climate cool, becoming warm, but no wetter than today. Early Middle Stone Age.
Bedrock weathered	*c.* 208 000–186 000	Significant chemical weathering prior to deposition of level 15.

Cave. The sequence begins at a time period estimated to be equivalent to oxygen isotope stage 5c/5d and below radiocarbon dates of greater than 48 500 BP (Pta-488) and carries through to deposits dated to 38 600±1500 BP (Pta-704) (Table 5.2). In the hiatus in human occupation in the sequence between about 38 000 and 1000 BP, at 13 300±150 BP (Pta-721) there is a brief period of (non-human) occupation with a sample of micromammalian bones (Beaumont *et al.* 1978; Avery 1982*a*). The micromammalian samples therefore give an indication of the range of small mammals collected by owls in the vicinity of the cave during much of the Last Interglacial/Glacial cycle, excluding the Last Glacial Maximum; there is a brief glimpse of conditions in the terminal Pleistocene, and modern conditions are represented in the samples dating within the last 1000 years (Table 5.2).

In the lowest unit in which micromammalian remains have been studied, assigned by Avery (1982*a*) to oxygen isotope stage 5c, the range of micromammalian species, their relative frequency and diversity

Table 5.2. Border Cave palaeoenvironmental inferences based on the micromammalian data (after Avery 1982*b*).

Excavation units	Approximate age (years BP)	Vegetation	General climatic conditions
1a	700 to present	Fairly open savannah woodland; some ranker vegetation	Very mild, dry
1b	*c.* 29–13 000	Open grassland on flats and slopes	Mild (late glacial), cool, dry
2–3	*c.* 38–33 000		Moderate to cool
4	*c.* 60–40 000		Mild, dry
5	*c.* 70–60 000	Thick dense grass probably fairly damp; more extensive forest or thick bush; scrub on slopes	Moderate
6–7a			Cool
7b–8	*c.* 100–80 000		Mild, relatively dry
9–10	>100 000		Cool, wet
11		Fairly open dry savannah woodland; some ranker vegetation	Mild, dry

suggest the vegetation was fairly open dry savannah woodland, with some rank vegetation along river courses. Cooler, moister conditions that probably led to thicker denser grass and more extensive forest or thick bush are assigned to the transition from oxygen isotope stage 4 and the beginning of stage 3. Stage 3 was generally moderate. Cool and dry to mild conditions prevailed at about 13 000 BP when the vegetation consisted of open grassland on flats and slopes. Within the last 1000 years, very mild and dry conditions with fairly open savannah woodland and some ranker vegetation prevailed under conditions that were somewhat warmer than those of the lowest samples in the sequence assigned to oxygen isotope stage 5c. In general, grassland forms predominant during the glacial and woodland forms were more common under interglacial conditions (Avery 1982*a*, p. 203).

Various parameters of the mandibles of *Crocidura flavescens* and *Crocidura hirta* were measured and the results show different directions of change in the two species. *Crocidura flavescens* is larger in the Holocene and smaller under glacial conditions providing an interesting but not easily explicable contrast to measurements of mandibles of the same species at Boomplaas Cave in the southern Cape where the opposite trend has been shown (Avery 1982*a*, p. 197). Changes in individual size therefore apparently cannot be related simply to temperature variation as has been assumed in the past (Avery 1982*a*).

In the Border Cave large mammal fauna, Klein (1977, 1980, p. 239) has shown that samples assigned to oxygen isotope stages 5c/d through 3 alternate between those in which bushpig, Cape buffalo, kudu, nyala, bushbuck and eland are more common, and those in which warthog, Burchell's zebra, and alcelaphine antelope predominate. Since the latter group were more prominent in the vicinity of the cave in historic times, it is assumed that the former group indicates periods when the vegetation was bushier than it is now (Klein *op. cit.*).

5.6.2 Other Transvaalian cave sites

Long sequence sites from which samples of larger mammals have been studied in the Transvaalian Ecozone include Border Cave (part of the Last Interglacial and the Last Glacial, and the end of the Holocene), Bushman Rock Shelter (the Last Glacial and the Early and Late Holocene), Heuningneskrans (the Last Glacial, Last Glacial Maximum and the Early and mid-Holocene), and Shongweni (Last Glacial Maximum and Late Holocene). Pomongwe in Zimbabwe probably dates to the Last Glacial, possibly the Last Glacial Maximum and the Holocene. The data have been summarized by Klein (1980).

The occurrence of the common springbok and blesbok/bontebok at the Cave of Hearths and Kalkbank in the Transvaal is interesting in that these finds are well beyond the northern limits of their ranges in historic times, suggesting that cooler conditions prevailed at the time the faunal samples accumulated in the early part of the Last Glacial (Klein 1980).

Faunal samples dating within the Last Glacial and the Last Glacial Maximum, however, show no variability, nor are there appreciable differences between them and Holocene samples from the same or nearby sites. For example, Heuningneskrans and the nearby Bushman Rock Shelter in the eastern Transvaal have the same range of species in their Late Pleistocene deposits as are found at Holocene sites nearby in Swaziland at Mlaula, Onink and Embabeni shelters (Klein 1984). Any differences that do occur can be explained as the result of variability in the local topography in the vicinity of individual sites. This demonstrates that, in contrast to the micromammalian evidence, and the predictions made for insects and small reptiles by Stuckenberg (1969), the larger mammals were apparently less affected by temperature and rainfall changes in the Late Pleistocene, although sample sizes are small and may account in part for the observed pattern. The ranges of a few ungulates were altered in the Transvaalian Ecozone, namely the common springbok and blesbok/bontebok, suggesting a considerable northward extension of grassland during the Last Glacial.

Three caves in dolomite bedrock in the eastern Transvaal have been investigated. In two of them, Heuningneskrans and Bushman Rock Shelter, Butzer (1984*a,b*) has studied the sediments built up in part through human occupation of the sites, and in the third, Echo Cave, changes in the water table have been linked to periods of travertine deposition (Brook 1982).

Heuningneskrans was formed by water emanating from a now defunct spring at the back of the cave and is situated close to the valley floor. The sediments give no clear indication of palaeoclimatic change over the past 32 000 years because the deposits have built up almost entirely through human occupation. Roof spall is almost absent (Butzer 1984*a*; p. 58).

The Bushman Rock Shelter sediments are deeper and probably date back beyond 60 000 years. Again no environmental changes could be detected in the upper levels spanning the Later Stone Age. There are two undated spall horizons in the lower part of the sequence, however, and these indicate cold conditions (Butzer 1984*a*; p. 59).

Echo Cave has no human occupation deposit, but travertine and sediments appear to record an oscillating water table that is assumed to have palaeoclimatic significance (Brook 1982). Two radiocarbon dates on travertine [32 115±965 and 31 210±1070 BP (no laboratory numbers given)] give a date for the oldest of three cold and wet periods. The two other phases when the water table was very high are undated, but Brook (1982) suggests they date, respectively, between 30 000 and 17 000 BP, and between 17 000 and 12 000 BP. A third radiocarbon date of 2675±75 BP is considered to date warm and dry conditions during the Holocene with short wet intervals. Without a more complete sequence of dates and temperature estimates, these are little more than general indicators of a higher water table at some stage between 30 000 and 3000 BP.

5.6.3 Palaeotemperature estimates from Wolkberg Cave

Major speleothem formation has occurred in underground chambers of the Wolkberg Cave south-east of Pietersburg in the Transvaal with columns more than 25 m in diameter and 30 m in height. At least two phases of increased solutional activity are evident from the sequence of speleothem deposition in deeper caverns (Talma *et al.* 1974). Samples were dated by radiocarbon and analysed for oxygen-18. Most of the stalagmites showed at least one re-solution phase. The average oxygen-18 content of two samples dated to 29 600 and 19 800 BP, when compared with the most recent samples from the Late Holocene, suggest a temperature depression of 8–9.5°C between the Last Glacial and the

Holocene. For this calculation, Talma *et al.* (1974, p. 139) have assumed, on the basis of data from the eastern Cape at Uitenhage (Heaton *et al.* 1986*a*), that the oxygen-18 content of cave seepage water during the Last Glacial was lower than during the Holocene. J. C. Vogel (personal communication) has dated 31 speleothems from dolomite caves in the northern, eastern, and central Transvaal by radiocarbon, and the results show that, except possibly for brief periods, there was always enough precipitation for the formation of cave sinter in the region during the Late Pleistocene and Holocene.

5.6.4 Colluvial and alluvial deposits in Swaziland

A sequence of alluvial and colluvial deposits with interstratified archaeological material is described by Price-Williams *et al.* (1982) from the Swaziland Middleveld. They recognize at least five phases of alluvial deposition in the region. The earliest are spreads of coarse gravel and pebbles which probably represent the residual of several periods of deposition in the Early Pleistocene. The Badzulu gravels contain archaic Earlier Stone Age artefacts, whilst the Lizembe gravels have an Acheulian industry. Along the Komati River, the finer gravels of the Shongololo alluvium contain Middle Stone Age material. The Winn alluvium is sandy in texture and archaeologically sterile, but overlies deposits containing MSA material and is overlain by Later Stone Age sites. The latest alluvium, the Yebo, is also sandy and occurs in the zone inundated by modern flood discharges.

Colluvial mantles consisting of poorly sorted intermixed clays, sands and fine gravels are widespread in the lower Middleveld between altitudes of 200 and 400 m, especially in areas of granite and granodiorite bedrock. The colluvia indicate slow deposition, as suggested by abundant interstratified unrolled artefacts, with long periods of stability marked by soil formation.

The lower colluvium is termed the Bovu and overlies fluvial gravels. It is up to 5 mm thick and was derived from pre-weathered granites. A period of considerable post-depositional stability is indicated by vermicular structures, probably termite tunnels or root casts, in its lower parts. The upper parts of the Bovu colluvium are iron enriched and have a deep red colour, with local ferricrete development. The final phase of the Bovu involved deep columnar cracking, later filled with calcium carbonate, and gully erosion. In some areas, a stone line, possibly equivalent in age to the Shongololo alluvium, divides the Bovu from the overlying Mphunga colluvium.

The Mphunga colluvium overlies the irregular upper surface of the Bovu and is well preserved over wide areas of Swaziland, and adjacent parts of Natal and Zululand where it has been mapped as the Estcourt soil series. The Mphunga has a greyish brown colour and is fine grained in texture. *In situ* working sites within the colluvium suggest episodic low energy deposition, an interpretation supported by a series of horizons of calcium carbonate nodules suggesting periods of syn-depositional soil formation. The upper parts of the Mphunga colluvium consist of a dark grey indurated horizon, which is locally dissected and everywhere covered by the unconsolidated sands of the present surface. The Mphunga is overlain by the Winn terrace sands and abutted by the recent Yebo alluvium.

Archaeological material and radiocarbon dates of 30 300±30 BP at the base and 8490±45 BP near the surface indicate that the Mphunga colluvium accumulated during the period 30 000–10 000 BP. Price-Williams *et al.* (1982) suggest that it results from geomorphic instability induced by a reduction in the vegetation cover as a result of a decrease in rainfall. The succeeding Winn alluvium was laid down in conditions of increased and possibly less seasonal rainfall in the Early Holocene.

Differences in the sediments and chemical characteristics of the Bovu and Mphunga colluvia are regarded by Price-Williams *et al.* (1982) as a consequence of the greater age of the Bovu colluvium, which has undergone a period of pedogenesis in warm and humid conditions prior to the later Pleistocene, although its age is not yet certain.

5.6.5 Coastal dune cordons and related deposits

The coastal plains of Natal and Zululand are covered by a series of Pleistocene dune cordons, between which are reworked dune sands. Locally, the sands overlie lagoonal muds and clays laid down in periods of high sea level. These deposits record the effects of Pleistocene sea level changes in some detail, but provide little in the way of palaeoclimatic information.

Maud (1968) identifies former sea levels at altitudes of up to 170 m above sea level. Those above 60 m are considered to be Lower Pleistocene, those at 30–60 m Middle Pleistocene, and those less than 30 m Upper Pleistocene and Holocene (SACS 1980).

Davies (1976) recognized one and locally two separate dune cordons in Natal and southern Zululand, and up to six cordons in northern Zululand, where they extend for up to 80 km inland. It appears that in Natal different generations of dunes are superimposed because of the narrow coastal plain. The older cordons are composed of decalcified red sands termed the Berea formation (SACS 1980), which overlie marine terraces and gravels containing Middle Pleistocene artefacts (Davies 1976). The younger dune cordons are, in contrast, composed of pale, partly or completely cemented calcareous aeolianites of inferred Middle to Upper Pleistocene age (Maud 1968; Davies 1976). These constitute the Bluff Sandstone Formation (SACS 1980). Controversy has surrounded the cause of the contrast between the red, decalcified Berea Formation and the calcareous Bluff Sandstone Formation. Maud (1968) felt that the difference was related to the degree of weathering and thus age, and that the calcareous dune cordons post-dated a Middle Pleistocene period of tropical weathering. Davies (1976) followed a similar line of reasoning, and concluded that the degree of reddening and decalcification was a function of the age of the cordons. However, Yaalon (1982) has argued that differences in the amount of calcium carbonate in the source materal have caused the contrast. The younger sands originally contained more than 25 per cent calcium carbonate which resulted in their cementation to form aeolianate. The parent material of the Berea sands had a low carbonate content which led to their decalcification and reddening. Davies (1976) regarded the dune cordons as having formed during the early stages of periods of regression, although the Bluff Sandstone Formation may have formed in part in a transgressive period (SACS 1980). Hobday (1976) inferred from cross-bedding dip directions that there was an increased frequency of onshore (south-easterly) winds during periods when the younger dune cordons were being formed.

Periods of low sea level were also accompanied by dune formation. Dunes which accumulated during a sea level at −68 m (possibly coeval with the Last Glacial Maximum) today form the Aliwal Shoal and Protea Banks (McCarthy 1967).

Between and underlying the dune cordons are lagoonal and estuarine deposits laid down in transgressive phases. The most extensive and best studied of these is the Port Durnford Formation (Hobday and Orme 1974; Hobday and Jackson 1979). This represents a complex of estuarine and lagoonal deposits, overlain by peats and barrier sands which lies beneath a coastal dune cordon. The deposits contain a rich mammalian fauna which includes elephants, hippopotami and other tropical species said by Hobday and Orme (1974) to be of Late Middle Pleistocene age. More recently, Hendey and Volman (1986) have identified a fossil pig, *Kolpochoerus*, amongst the Port Durnford fauna that dates to the Early Pleistocene. For this reason they have challenged the earlier assumption that the deposits mark the rise of sea level at the Last Interglacial, and suggest instead that the maximum level of +8 m relates to an Early Pleistocene transgression. The lagoonal and barrier facies contain tropical mollusca, and the dune cordons were formed in subsequent regressions.

Deposits related to the Last Interglacial transgression include fossiliferous limestones at Lake St Lucia (Hobday 1976) at a sea level +3.5 m or more higher than today. The fossils include corals and thermophile mollusca which suggest water temperatures much higher than those of today. Characterizing this transgression is the distinctive Swartkops fauna (Davies 1980) which includes many tropical and sub-tropical species of mollusca.

During the low sea levels of the Last Glacial Maximum, all the major rivers of Natal and Zululand excavated deep valleys to depths of −67 m (Maud 1968). Their valleys were infilled intermittently with intercalating marine, estuarine, and fluvial deposits during the Holocene rise in sea level (Orme 1976).

5.6.6 Wonderkrater and other pollen sequences

The longest pollen sequence in the Transvaalian Ecozone is that studied from organic thermal spring deposits cored at Wonderkrater in the west-central Transvaal which falls within the better-watered eastern part of the Transvaalian Ecozone. The sequence, which has 12 radiocarbon dates, covers the last 35 000 years or more and has been divided into 10 pollen zones from W1 at the base to W10 at the top (Scott 1982*a*). At present, the area receives between 400 and 600 mm precipitation per annum and has on average 5–10 days of frost a year. Although the wider surroundings of the site are typical of the *Combretum* veld type of the plains bushveld, the savannah around the spring site has more microphyllous plants associated with the local soil and spring waters. In order to aid in the evaluation of the fossil pollen spectra, modern surface pollen samples from 21 Transvaal, and four other sites were examined (Scott 1982*a*, pp. 351–4; Fig. 5.8). Because some of the radiocarbon samples appear to have been contaminated and have given dates that are out of chronological order in the stratigraphic succession, estimates have been calculated and are used in the following description of the sequence (Scott 1982*a*; Table 5.3).

The base of the sequence, pollen zone W1, is at least 35 000 years old. It has a high percentage of arboreal pollen (AP) with up to 34 per cent *Podocarpus*. The reconstructed vegetation indicates a cooler temperature (mesothermal), moist subhumid climate that was probably slightly warmer and drier at the beginning of the zone than at the end.

Pollen zone W2 covers the time period between 20 000 and 16 000 BP. Arboreal pollen decreases with *Podocarpus* represented by around 10 per cent and there is a relatively high percentage of grasses, particularly in the lower levels, as well as macchia elements. The climate is suggested to have been cooler temperate (mesothermal), dry sub-humid or semi-arid, much drier than during pollen zone W1.

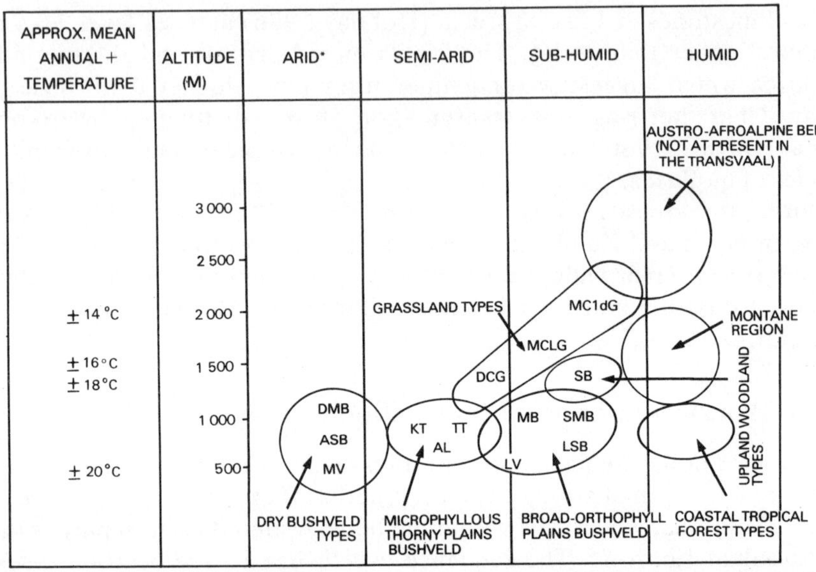

Fig. 5.8 Schematic representation of the relationship between Transvaal vegetation types and altitudinal and moisture gradients. Veld types after Acocks (1953). + Estimated from Schulze and McGee (1978); * after Poynton (1971). Diagram from Scott (1982*a*).

A cooler climate, but one not as arid as W2, is represented by the pollen spectrum in zone W3 which pre-dates 15 000 BP. Forest elements are still relatively rare with the AP constituting less than 25 per cent of the pollen sum in W3 and in W4 which pre-dates 11 000 BP. The reconstructed vegetation of W3 and W4 suggests cold conditions possibly approaching those of an ericaceous vegetation type, and a slight amelioration of temperatures in the upper part of the zone.

Pollen zone W5 has an interpolated age of between 11 000 and 9500 BP. The AP percentage is still relatively low (±12 per cent), but with a peak in the lower portion of one of the two boreholes covering this pollen zone of about 42 per cent indicating some variability. Grassland and shrub taxa have become more common than Ericaceae, and the reconstructed climate suggests conditions that were warmer and drier than pollen zone W4, and slightly cooler than those of today.

Zone W6 probably covers the time period between 9500 and 6000 BP. There is sharp increase in the AP percentage which ranges between 33 and 56 per cent of the pollen sum. Bushveld vegetation resembling Kalahari thornveld predominates suggesting temperatures warmer than during W5 and somewhat drier than at present. Between 6000 and 4000 BP, the pollen spectra from zone W7 indicate somewhat cooler temperatures with a rise in moisture until at the top of this zone moisture

Table 5.3. Wonderkrater summary of vegetation and climatic inferences based on pollen analyses (after Scott 1982*a*).

Zone	Vegetation	Climate	Approximate age (years BP)
W10	Present broad orthophyll or microphyllous-type bushveld with restricted montane forests	Warmer temperate, dry? subhumid	1000 to present
W9	Broad orthophyll plains or upland bushveld with restricted montane forests	Warmer temperate, subhumid	*c.* 2000–1000
W8	Upland bushveld type with restricted montane forests	Cooler temperate, subhumid	*c.* 4000–2000
W7	Savannah with broad-leafed elements with montane forests virtually absent	Warmer temperate, dry subhumid	*c.* 6000–4000
W6	Kalahari-type savannah with montane forests absent	Warmer temperate, semiarid	*c.* 9500–6000
W5	Open grassland with much reduced, virtually absent montane forests	Cooler temperate, semiarid	*c.* 11 000–9500
W4	Open grassland with ericaceous belt elements and restricted montane forests	Cooler temperate to microthermal, subhumid?	*c.* 25 000–11 000
W3	Open grassland with more ? *Podocarpus* forests	Cooler temperate? subhumid?	? *c.* 25 000
W2	Kalahari-type savannah with restricted montane forests	Cooler temperate, semiarid, or dry subhumid	<34 400
W1	Relatively mesic woodland with expanded montane forests	Cooler temperate, moist subhumid	*c.* 34 400+

levels were about the same as those of today. These somewhat cooler temperatures persisted through to zone W8 which accumulated between about 4000 and 2000 years ago. The vegetation suggests open savannah.

Pollen zone W9, dating to about 2000 BP, shows an increase in AP up to 42 per cent indicating a denser bushveld vegetation suggesting warmer temperate, subhumid conditions that were not as warm nor as dry as those of the present. At the top of the sequence in zone W10, conditions of the last 1000 years were not very different from those of today, but possibly slightly more mesic.

Also in the Transvaal, at Rietvlei dam on the highveld, Scott has identified five pollen zones dating from about 10 300 to about 1300 BP (Scott and Vogel 1983). At the base of the sequence more open grassland vegetation indicates cooler conditions than those of the present that were replaced by more arid elements up to about 7500 BP. Thereafter, bushveld expanded with warmer and moister conditions with a climate and vegetation similar to that of the present after about 3000 BP. A spring deposit at Scot in the northern Transvaal Lowveld has bushveld pollen spectra throughout the last 5000 years, with lower percentages of AP between *c.* 3000 and 2000 BP suggesting somewhat cooler temperatures at that time (Scott 1982*b*). Another dated spectrum, from the Moreletta River near Pretoria, confirms more open vegetation and cooler conditions prior to 7500 BP, a wetter and warmer climate around 5200 BP, and slightly drier or cooler conditions than the present around 440 years ago. This profile does not, however, provide clear evidence for optimum temperatures during the Holocene although this may be a result of a break in sedimentation within the sequence (Scott 1984*b*).

An undated series of pollen samples from an alluvial terrace and a spring on the farms Theuniskloof and Tugela along the Limpopo River in the northern Transvaal indicates cooler conditions with more open bushveld than at present which could date either to the end of the Pleistocene or to the early Holocene (Scott 1984*b*).

The Holocene sequence noted south of the Limpopo with more open grassland in the Early Holocene and bushier environments in the mid-Late Holocene is repeated in a pollen sequence from the Inyanga Highlands in Zimbabwe that dates from about 12 000 BP to modern times (Tomlinson 1974; Scott 1984*a*).

5.6.7 Charcoal analyses

There is some potential for the use of charcoal analysis to check the pollen record in the Transvaalian Ecozone. One study has so far been undertaken at Siphiso rock shelter in north-east Swaziland (Prior 1984). Samples from Holocene deposits about 2 m deep show wetter conditions between *c.* 9000 and 6000 BP, a drier mid-Holocene, and moister conditions again in the last 4000–3000 years (Prior and Price-Williams 1985).

5.6.8 Tree rings

In a pilot study of tree-rings records in the Transvaal, Gillooly (1975) has shown that the quasi-20-year cycle of drier and wetter years observed in rainfall records over the last 100 years is confirmed in the tree ring data that are currently available. There is clearly some potential in this method if older trees can be found to extend the record further back in time.

5.6.9 Planktonic foraminifera

Two short gravity cores have been examined for evidence of past ocean temperatures from the southern Mozambique Channel on the lower continental slope dipping from the Mozambique Plateau to the Natal Basin at depths of 2750 and 3800 m, respectively (Vincent 1972). The two cores show similar trends in the distribution of planktonic foraminiferal species, the older parts being characterized by an abundance of temperate-water species and the more recent parts showing a marked increase in warm-water species. There is also a shift in the coiling direction of populations of *Globorotalia truncatulinoides* from dominantly dextral through most of the core to dominantly sinistral at the top. Finally, there is a change in the abundance of radiolarians in both cores, the number increasing by several orders of magnitude after 10 000 BP. Radiocarbon dates (which should be considered as minimum determinations) range from 28 940±5500 BP at the base to 536±1724 BP at the top, with the major faunal changes dating to about 10 000 BP (Vincent 1972).

The changes are interpreted as reflecting a warming of about 5°C in the surface waters of the southern Mozambique Channel that is contemporaneous with a similar warming in waters of the south-east Indian ocean at about the same latitude (Vincent 1972).

5.6.10 Archaeological proxy evidence

With the exception of sites in the Matopos Hills of Zimbabwe, very few archaeological deposits at caves and rock shelters in the Transvaalian Ecozone date within the period 9000–4500 BP. The reason for the much higher incidence of Early and Late Holocene occupation is obscure; it was initially suggested (Van Zinderen Bakker and Butzer 1973; J. Deacon 1974) that warmer conditions in the mid-Holocene after 6000 BP may have been accompanied by lower rainfall making it more difficult for hunter-gatherers to find a reliable range of plant foods, but the palynological research summarized above indicates rather that the mid-Holocene in the Transvaal was warmer and slightly moister than the Early Holocene. This pattern of human occupation clearly needs further investigation and a larger sample of sites and excavations at Jubilee Shelter (Wadley and Verhagen 1986) have shown that there was mid-Holocene occupation of at least some sites in the Transvaal.

5.7 Chronological summary for the Transvaalian Ecozone

Where longer sequences are available in this ecozone, generally cooler temperatures with warmer intervals can be identified in the cave sediments and micromammalian data dating to the end of the Last Interglacial and the beginning of the Last Glacial, with generally moderate conditions prior to the Last Glacial Maximum. While the details are much less obvious in the larger mammal faunas, they too indicate generally cooler conditions during the Last Glacial with several relatively warmer intervals. The longest pollen sequence is no older than about 35 000 BP, but is useful in including evidence for wetter conditions between *c.* 35 000–25 000 BP, and Last Glacial Maximum conditions showing a decrease in trees accompanied by an increase in grasses and macchia elements. A similar decrease in tree cover is seen in the late glacial micromammalian sample from Border Cave suggesting it was cold and dry. The Holocene pollen sequences suggest warmer and relatively dry conditions in the Early Holocene with grassland prevailing and warmer and slightly moister conditions in the mid-Late Holocene when environments were bushier. The charcoal evidence from Siphiso does not always agree with the Wonderkrater sequence, showing rather moister conditions in the Early Holocene, a drier mid-Holocene, and a return to moister conditions in the Late Holocene. The deep-sea cores date the post-glacial warming of Indian Ocean waters to a minimum of 10 000 BP.

5.8 The Basutolian Ecozone

The Basutolian Ecozone (Fig. 5.9) comprises the mountains of Lesotho and the Drakensberg escarpment, together with the adjacent plateau areas in the Orange Free State, southern Transvaal, and the north-eastern Cape, which lie at altitudes of 1200–1800 m. The climate varies from sub-humid along the escarpment areas to the east where mean annual rainfall totals are around 1000 mm, to semi-arid in the west (mean annual rainfall 300–400 mm). Summers are warm and moist, but winters are cold and dry. Drainage on the plateau areas is centred on the Vaal–Orange system which drains to the Atlantic. Eastern and escarpment areas drain towards the Indian Ocean. Over wide areas of the plateau, pans are common between the widely spaced river valleys.

On the highveld to the west of the Drakensberg mountains, the vegetation in historic times was grassland, with trees confined to the river valleys. On the eastern escarpment zone, the grassland was interspersed with temperate forest in valley heads, with acacia savannah at lower altitudes. The larger mammals in historic times were mostly large gregarious grazers such as Burchell's zebra, quagga, black wildebeest, blesbok, and springbok which migrated seasonally with the availability of grazing. Browsers were rare, but mixed feeders such as eland and

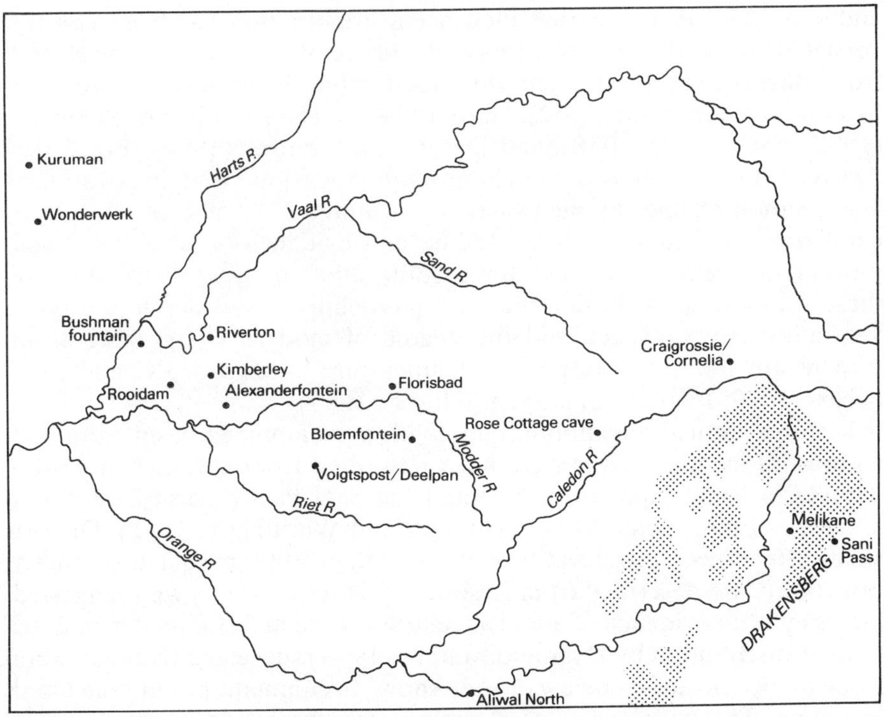

Fig. 5.9 Location of sites in the Basutolian Ecozone. Sites in East Griqualand are marked on Fig. 6.1.

steenbok were widespread, and warthogs were more common than bushpig (Klein 1980).

As elsewhere in Southern Africa, depositional basins are rare in this ecozone. Important areas where deposition has occurred include the alluvial sequences of the lower Vaal, Orange and Riet rivers; pans in the Orange Free State, thermal springs at Aliwal North and Florisbad, and peat bogs in Lesotho. These latter deposits have yielded important pollen sequences. In common with the other ecozones, archaeological cave deposits provide the most complete sequences, and the Basutolian Ecozone is unique in possessing periglacial features that are direct evidence of colder temperatures in the Late Pleistocene.

5.8.1 Periglacial features in the highlands of Lesotho

The mountains of Lesotho and adjacent areas of Natal rise to an elevation of over 3000 m, and form the largest and highest mountain mass in Southern Africa. In common with other high altitude areas in subtropical latitudes, the recognition and extent of landforms, deposits, and soils related to possible Pleistocene frost action has been controversial (Butzer 1973*b*). A variety of apparently active periglacial and frost phenomena have been described from high altitude areas of Lesotho by Alexandre (1962), Harper (1969), Hastenrath and Wilkinson (1972), Fitzpatrick (1978), and Dardis and Granger (1986) (Fig. 5.10). Active features, such as patterned ground, downslope sludging of stones, and ablation of fines by meltwater occur above 3100 m; pipkraken, turf exfoliation, and thurfur above 280 m, and evidence of accelerated soil movement occur on slopes above 1800–2000 m. The distribution of these forms reflects the interplay of microclimate, soil depth and type, vegetation cover, aspect, and the degree of modern overgrazing in an area where today the July mean temperature is around 3°C and frost occurs on 90–150 days of the year (Harper 1969).

Features termed 'nivation cirques' with a pronounced basin form and an infill of angular debris have been described from altitudes of 1600–2100 m in Natal (Sparrow 1967) and the Sani Pass area of Lesotho at altitudes of more than 2900 m (Marker and Whittington 1971). Further smaller features with elevations of more than 3000 m and a southerly orientation are described from Lesotho by Harper (1969) who suggested that they were evidence of a Pleistocene snowline at 3300 m. An analysis of their distribution by Dyer and Marker (1979) suggested that they were formed by nivation below local snow accumulations in sheltered locations. The wide range of elevations for these features and others

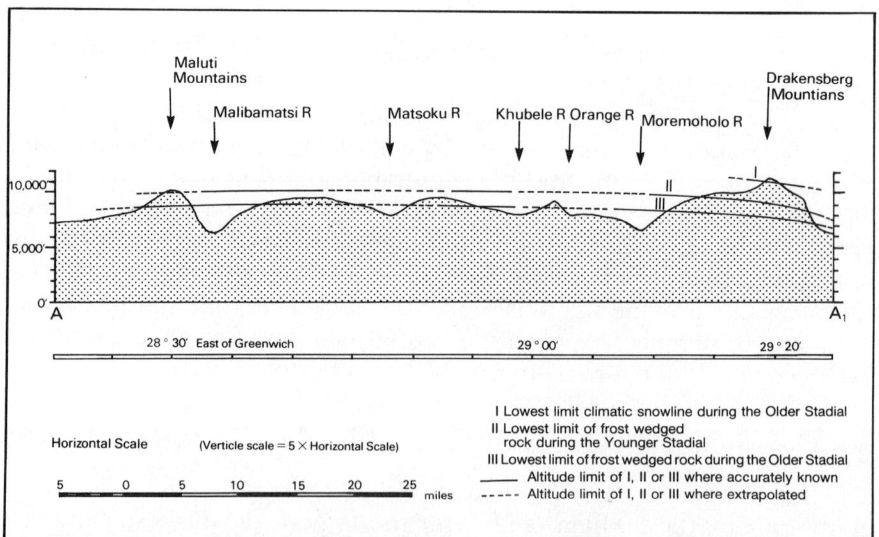

Fig. 5.10 Altitudinal zonation of periglacial features in Lesotho (after Harper 1969). (Reprinted from: *Palaeoecology of Africa*, (ed. J. A. Coetzee and E. M. van Zinderen Bakker) Vol. 4, 1967. A. A. Balkema, P.O. Box 1675, Rotterdam, Netherlands.

which suggest localized snow patch erosion were rationalized by Butzer (1973*b*) who noted that snowlines would be lower on windward (Natal) slopes compared to leeward (Lesotho) slopes. These features may also be of a variety of ages.

Other 'fossil' periglacial features are easier to interpret. Sparrow (1967, 1971) and Harper (1969) describe three different generations of blockfields, together with solifluction lobes and very large boulders in colluvial and alluvial deposits from the Natal Drakensberg at elevations above 1500 m in the south to 1700 m in the north. Two generations of over-steepened, south-facing slopes resulting from selective frost action on weaker basaltic lava flows were recognized by Harper (1969). The older features are common in Natal above 2100–2350 m, whilst the younger only occur above 2900 m. Fitzpatrick (1978) described a distinctive soil profile with a compacted horizon which occurs in areas above 2280 m in Natal. By analogy with similar features found in high latitude areas, such horizons form by slow freeze-thaw cycles in a perennially frozen soil, conditions which possibly existed at this altitude in Southern Africa during the Last Glacial cycle.

Harper (1969) calculated that a 9°C temperature lowering would have been required to produce the older fossil periglacial features and a 5°C temperature lowering to produce the younger features. Although Butzer (1973*b*) questions the assumptions on which these inferences were made, the temperature depression calculated for the younger features is consistent with the 5°C temperature depression generally accepted for the Last Glacial Maximum (CLIMAP 1976) as well as that derived from isotope studies by Heaton *et al.* (1986*a*) for the Last Glacial Maximum period at lower elevations in the eastern Cape. Talma *et al.* (1974) suggest a temperature depression between 8 and 9.5°C for the period 30 000–20 000 BP and the older features could have been formed in this period when temperatures in the highland interior of Southern Africa may have been 8–9.5°C colder than at present (Talma *et al.* 1974).

Dating the periods of periglacial activity is difficult, as noted by Sparrow (1967, 1971) and Harper (1969). Many of the features may well be older than the Late Pleistocene (Butzer 1973*b*) and Ellenberger (1960) notes that the older generation of solifluction grades to pre-Acheulian (Earlier Stone Age) river terraces. Visser and Van Riet Lowe (1955) and Butzer (1971) note that screes pass into colluvium which is associated with Late Pleistocene 'older fills'.

Carter (1976) describes three periods of intensified frost action in Lesotho as evidenced by increased spalling of material from the roof of Melikane Cave. On the basis of bracketing radiocarbon dates and associated Middle Stone Age artefacts, these periods occurred before 43 000 and possibly after 65 000 BP, between 42 000 and 35 830 BP, and between 32 200 and 30 900 BP. There appears to be a correlation between the youngest of these periods and the periods of greater cold and frost action noted at Wonderwerk and Rose Cottage caves by Butzer (1984*b*) and on the Gaap escarpment by Butzer *et al.* (1978*b*).

Vogel (1983) measured the carbon-13 content of collagen in a series of zebra teeth from Melikane Cave which showed that their diet included about 35 per cent C_3 plants in the Late Holocene, but between 75 and 84 per cent C_3 plants between 42 000 and 20 000 BP (Table 5.4). This indicates a clear shift in the distribution of C_3 plants as a response to lower temperatures during the Last Glacial.

Table 5.4. Carbon-13 content of collagen from *Equus* teeth from Late Pleistocene and Late Holocene levels excavated at Melikane Cave in Lesotho by Carter (1976). The last column lists the contribution of C_3 plants to the diet of the animals as deduced from the isotopic composition. The results clearly show a different composition of C_3 and C_4 plants in the vegetation cover during the Late Pleistocene (after Vogel 1983).

Sample	Approx. age (years BP)	Carbon-13	% C_3 in diet
MLK2	1500	−12.8	38
MLK3	1500	−12.2	34
MLK9	20 000	−18.0	75
MLK10	20 000	−18.3	77
MLK12	20 000	−18.5	79
MLK13	35 000	−19.2	84
MLK14	35 000	−19.2	84
MLK20	42 000	−18.8	81

Little evidence is available for changes in moisture conditions in the highlands. A pollen profile from an alpine bog at Mekhaaba at the Madibamatso riverhead in Lesotho was studied by Van Zinderen Bakker (1955). Although there are no radiocarbon dates from the profile, Van Zinderen Bakker and Werger (1974) note that other bogs in the region all date to within the last 8000 years. A gradual decline in the percentage of Cyperaceae, offset by an increase in Gramineae and Compositae in the spectra from the two pollen zones, suggests that the climate became drier during the Holocene (Scott 1984*a*, p. 73).

5.8.2 The subhumid eastern plateau and escarpment zone

The eastern parts of the Orange Free State and the north-eastern Cape

Province, together with adjacent areas of Natal, experience a subhumid climate with an annual rainfall of 500–1000 mm.

Aliwal North and Florisbad

The Aliwal North sequence (Table 5.5) was studied by Coetzee (1967). It is of particular importance because the site is near the ecotone between the dry Karoo and the somewhat moister *Cymbopogon-Themeda* grassland of the Basutolian Ecozone. Pollen spectra from the base of the sequence with an estimated age of more than 13 200 BP contain high proportions of *Stoebe, Typha,* and *Cyperaceae*, and indicate that conditions were cooler and more humid than at present. The presence of *Stoebe* indicates that the grassland had a composition quite different from that occurring in the area today (Scott 1984*a*). During the following 3000–4000 years, pure grassland alternated with Karroid vegetation several times, indicating that both cooler and moister, and warmer and drier intervals occurred. An abundance of Karroid vegetation in pollen spectra dated to 11 650–11 250 BP was regarded by Coetzee (1967) as indicating a warm dry phase coeval with the Allerod phase of the European late glacial sequence.

Table 5.5. Inferred vegetation and climatic changes from pollen samples from Aliwal North (after Coetzee 1967).

Pollen zones	Radiocarbon (years BP)	Vegetation	Inferred climate
Zii	9650±150	Karroid	Warm and dry
Zi		Dry grassveld	Becoming warmer and drier
Y		Grassveld	Cooler and moister
X	11 250±180 11 650±170	Karoo vegetation	Very warm and dry
W	12 200±180	Pure grassveld	Cooler and moister
V		Karroid	Warm and dry
U	12 600±110	Pure grassveld with *Stoebe plumosa*	Cooler and humid

The thermal spring at Florisbad has a 7-m high mound of spring sands with interbedded organic horizons, some of which contain pollen. When first described by Van Zinderen Bakker (1957), the radiocarbon dates suggested that the pollen bearing part of the sequence was of Holocene age. A second sequence of dates suggested that the upper part of the sequence at least, dated to the period between 29 000 and 19 000 BP (Van Zinderen Bakker 1967, 1976), but it has now been established that most of the pollen bearing sequence is beyond the range of radiocarbon dating and is at least 35 000 years old (Kuman and Clarke 1986). The sediment column below Peat II is of Late Middle Pleistocene age and includes six extinct mammalian species or genera (Klein 1984).

Sediment analyses (Butzer 1984*a*) show that the sequence consist of several cycles of spring activity which appear to represent changes in groundwater discharge. Each cycle is marked by a rapid coarsening of sands which then become finer after a period of strong, but fluctuating discharge until accretion of sands is minimal and organic matter begins to accumulate around the relatively stable spring. At this time, vegetation growth and moisture around the spring is at a maximum, as suggested by the occurrence of aquatic mammals such as hippopotamus, lechwe, water mongoose, and clawless otter in Peat I. The occurrence of giraffe indicates the presence of trees in a region where they were virtually absent in historical and modern periods, yet there is no arboreal pollen in any of the deposits from the site (Klein 1980).

During the Holocene, there were two periods of greater spring discharge, the first of which occurred 10 000–5500 BP. They were paralleled by moderate energy alluviation on the nearby Modder River (Butzer 1984*a*). Butzer (1984*a*) suggests that Peat III formed between 28 500 and 18 400 BP. The pollen associated with this peat indicate a considerable increase in the grass cover, and cooler and/or wetter conditions (Van Zinderen Bakker 1976), although it is not possible to ascertain whether this grassland was equivalent to the *Cymbopogon-Themeda* grassland of the plateau and eastern Orange Free State, or the Austro-Alpine grassland found at higher altitudes (Coetzee 1967; Van Zinderen Bakker 1967).

The early part of the sequence dates to the Middle or early Late Pleistocene (Kuman and Clarke 1986). Very high spring discharge followed the accumulation of Peat II and suggests intense periodic rains. The pollen spectra indicate a dry climate with Karroid vegetation in which Compositae are better represented than they are in modern pollen spectra (Van Zinderen Bakker 1957). This was gradually replaced by grassland, with locally wet habitats, indicated by the fauna, which again includes lechwe (Klein 1980).

The Little Caledon Valley

Terrace gravels and sands, some containing artefacts, were described from the narrow valley of the Little Caledon River south of Bethlehem by Visser and Van Riet Lowe (1955). They identifiy an older pre-Late Pleistocene suite of gravels with discontinuous terrace levels, and a younger suite of fine grained deposits with sands and silts up to 15 m above the river level. Later Stone Age artefacts are reported from the surface of the highest terrace. Visser and Van Riet Lowe interpreted the gravels as evidence for wetter climates with the sands being deposited in periods of lesser, but still increased, rainfall. There are clear parallels between this sequence, and the Younger Gravels of the Vaal River (Rietputs Formation) and the Riverton Formation of the Lower Vaal. The sediments and stratigraphy of these deposits clearly warrant further investigation.

Pollen spectra from Cornelia and Craigrossie, which are situated at elevations of 1700–1800 m in the Clarens area of the foothills of the Lesotho highlands were studied by Scott (1986). Vegetation in the study area today is grassland, with a shrubby ericoid element on mountain tops above 2000 m and some montane woodland in protected and south facing ravines. Scott (1986) tentatively recognized seven distinct pollen zones, spanning the last 20 000 years (Table 5.6). At Craigrossie, the lower part of the sequence overlaps that from Aliwal North. Palaeoclimatic interpretations of the pollen spectra are difficult and the radiocarbon dates are not all in stratigraphic order. It appears that conditions prior to 12 600 BP, possibly back to *c.* 20 000 BP, were relatively cool and dry, evidenced by an increase in ericaceous pollen,

Table 5.6. Cornelia and Craigrossie: summary of palaeoenvironmental inferences based on pollen analyses (after Scott 1986).

Approx. age (years BP)	Lithology	Zone	Vegetation	Climate
Craigrossie:				
Historical	Grey and red sand	CR 5	Grassland, locally moist	Relatively moist, dry winters
c. 660	Organic loam			
Holocene	Pale sand	CR 4	Grassland with Compositae	
c. 10 700	Grey sand Brown silt	CR 3	Grassland	Less dry than CR 2, becoming wetter
	Sandy loam interval	CR 2	Karoid grassland with local Chenopodiaceae	Relatively dry. Dry winters?
	Brown silt	CR 1	Grassland with fynbos. Distant forest expansion	Relatively moist. Less winter drought?
<12 600				
Cornelia:				
c. 12 600	Black silt Grey sand	C 2	Fynbos with local sedges	Relatively cool moist. Less winter drought?
c. 20 000	Silt	C 1	Fynbos with bryophytes	Relatively moist but cooler and/or drier than C 2

sedges and bryophytes, but without a parallel increase in arboreal pollen. Between 12 600 and 10 700 BP, an increase of arboreal pollen, especially *Podocarpus*, indicates an expansion of forests in mountain valleys and suggests that temperatures rose to close to present values. An increased proportion of shrubby vegetation, compared with the Late Holocene, suggests that there was both summer and winter rainfall. A change to drier conditions later in this period and an increase in summer rainfall intensity is marked by deposition of sandy and pebbly sediments, and an increase in Chenopodiaceae pollen, coeval with the drier conditions noted from Aliwal North. Holocene pollen spectra suggest an increase in available moisture and the development of a grassland, with Compositae and, locally, *Artemisia*. In the Late Holocene, an increased abundance of semi-aquatic elements in the pollen spectra suggest relatively moist local conditions. This may be correlated with the development of hydromorphic floodplains after 4000 BP as noted by Butzer (1984*a*) from the Welgevonden area.

Rose Cottage Cave

Rose Cottage Cave is a large sandstone overhang at an elevation of 1600 m near Ladybrand in the eastern Orange Free State. Sediments have accumulated behind a large fallen roof block. The sediments consist of fine and coarse roof debris derived from granular weathering and roof spall, together with material washed into the cave by upland drainage waters that have intermittently entered the back of the cave through a major joint intersection and drained out past the fallen block (Butzer 1984*a*). Often, the waterlain sediments are well laminated, and indicate heavy and protracted rains and, at least, a seasonally high water table. Clay and silt rich strata indicate low energy spring influx, and less intense rainfall and run-off. Angular roof spall is derived by frost weathering. The archaeological sequence ranges from Middle Stone Age (Howiesons Poort stage) to the most recent part of the Later Stone Age (Beaumont and Vogel 1972). However, the occupation has not been continuous.

Near the base of the occupation sequence associated with a Howiesons Poort assemblage sediments are interpreted as indicating cold conditions and prior to 50 000 BP, in a late Middle Stone Age level, spring activity, and roof spalling suggests a higher water table, and moister and cool to cold conditions. Late periods of increased spring activity and thus higher local rainfall appear to have occurred around 29 000 BP, between 8640 and 6850 BP, and again around 1100 BP. Unit RC-10C contains a major subangular to angular roof spall, as well as evidence for low spring inflow and a low water table, suggesting to Butzer (1984*a*, p. 57) that conditions were cold and relatively dry around the Last Glacial Maximum.

After a break in sedimentation, Later Stone Age people occupied the site between *c.* 8640 and 6850 BP when the water table was seasonally high, and again in the last 1100 years when the water table was high.

5.8.3 Eastern Griqualand

Analyses of charcoals from rock shelter sequences in this area by Tusenius (Opperman 1982; Tusenius 1986) provide important information on climate and vegetation changes across the Pleistocene–Holocene boundary (Fig. 5.11).

Charcoals from Bonawe, which lies below the escarpment, show that there has been no significant change in the woody vegetation over the past 8000 years. There is some evidence that the diversity of woody species was greater after 3000 BP, but the Early Holocene sample is very small. *Olea* spp., *Rhus* spp., *Acacia*, and *Maytenus/Pterocelastrus* are present throughout the sequence. In the later Holocene, *Euclea/Diospyros*, cf. *Clausa*, cf. *Celtis*, and four unidentified taxa were added to the range of firewoods utilized. Drier conditions are indicated for the period around 2250 BP by a significant decrease in the size and number of vessels in charcoals of *Rhus* and *Olea* compared to those in samples dated to 8040± BP (Tusenius 1986, table 12).

At Ravenscraig and Colwinton, both above the escarpment, changes in climates around 10 000 BP are evident. The basal deposits at Ravenscraig, which date to 10 000–12 000 BP, contain a relatively high percentage of *Leucosidea sericea* which favours relatively moist soil conditions. The overlying stratum, which has a radiocarbon date of 10 000 BP, is dominated by *Euryops* spp. and *L. sericea* constitutes less than 2 per cent of the charcoal samples. *Euryops* continues to dominate the mid-Holocene samples, but after 3000 BP is much rarer when *Cliffortia* spp. and *L. sericea* return to prominence. At Colwinton, the oldest sample dated to 6720 BP, has relatively high percentages of both *Euryops* and *Cliffortia* spp. However, in samples dating to 1890±70 BP *Euryops* is of minor importance and *Cliffortia* dominates the assemblage. There are indications of drier conditions in the Early and mid-Holocene based on spiral thickening of cell walls in some taxa. An increased percentage of alcelaphines and equids in Early Holocene deposits at Colwinton suggest warmer conditions at this time (Opperman 1982), although at Bonawe, below the escarpment, the Late Holocene fauna includes the alcelaphines. A similar observation was made at Good Hope Shelter in southern Natal where a very small sample of larger mammals consists only of alcelaphines and an equid in the Early Holocene (Cable *et al.* 1980).

As a regional sequence, the data from these sites indicate cooler and moister conditions prior to 10 000 BP with warmer and drier climates to

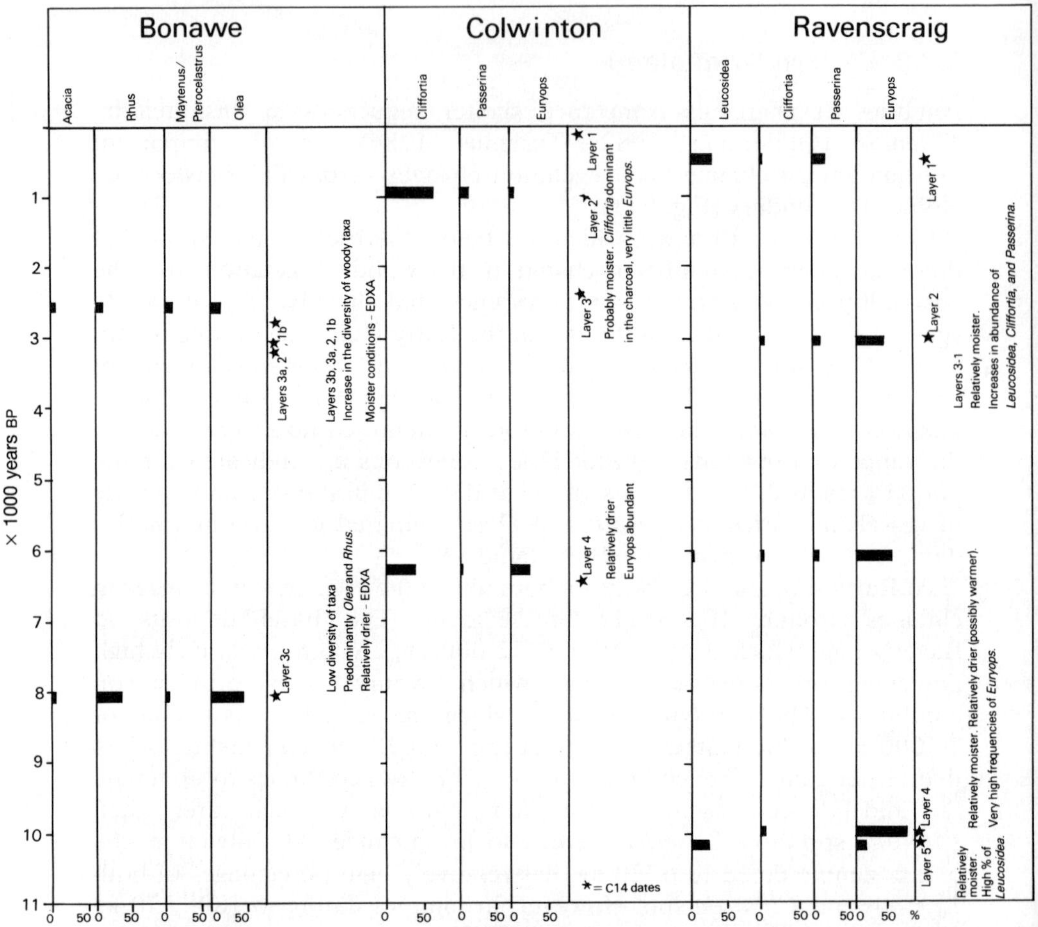

Fig. 5.11 Percentage frequencies of woody plant species identified from charcoals at Bonawe, Colwinton and Ravenscraig, with a summary of general climatic trends (after Tusenius 1986).

about 3000 BP. Thereafter, it was slightly cooler and moister again (Fig. 5.11).

5.8.4 The semi-arid zone

The western parts of the Basutolian Ecozone have a semi-arid climate with annual rainfall totals of 300–350 mm. The seasonal distribution of rainfall varies from south to north. In the south, 75 per cent of the total falls in the winter, whilst in the north-east over 80 per cent falls in the summer.

Alluvial deposits of the lower Vaal and Riet rivers

The alluvial deposits of these rivers constitute one of the few extensive areas of Late Cainozoic deposition in the Basutolian Ecozone. The pioneering studies of Sohnge *et al.* (1937) linked the sequences of alluvial deposits they observed to the then current East African pluvial-interpluvial scheme. Subsequent studies by Partridge and Brink (1967), Butzer *et al.* (1973*b*) and Helgren (1979) have shown that the sequence identified by Sohnge *et al.* (1937) is more complex than they envisaged and that the age of the deposits is considerable. The Older Gravels are of Middle to Late Tertiary age (Helgren 1977) on the basis of faunal correlations, whilst the Younger Gravels [the Rietputs Formation of Helgren (1979)] are apparently of Lower to Middle Pleistocene age. Only the Riverton Formation includes deposits that are of Late Pleistocene age.

The Riverton Formation is divided into five units, all of which are basically low to moderate energy floodplain overbank silts and sands. Members I and II contain Acheulian artefacts indicating a later Middle Pleistocene age, based on correlations with the Rooidam site dated by thorium/uranium to 174 000 BP (Szabo and Butzer 1979). Member II, which represents a complex aggradation of sands derived from the erosion of valley margins, lies beyond the range of radiocarbon dating and was probably deposited during Last Interglacial times (Butzer 1984*a*). It is coeval with Middle Stone Age settlement in the area. Considerable surface run-off and moderate stream energy are implied, the result of intense summer rains and an incomplete vegetation cover caused by an intensified dry season, in the period before 37 500 BP. A long interval of time separates the deposition of units III and IV during which pedogenic calcretes developed in the sediments of unit III and rivers incised by 8–15 m (Butzer 1984*a*). Radiocarbon dates on the calcretes suggest that they formed between *c.* 18 600 and 15 800 BP.

Member IV has two distinct sub-units, formed in different climatic conditions and separated by downcutting. Sub-unit IVa consists of sands, with occasional pebble lenses, derived from massive erosion by surface run-off of reworked weathered aeolian materials and colluvium (the Hutton sands). Valleys were choked with sediment and Butzer (1984*a*) suggests that a marked deterioration of the vegetation cover and drier conditions occurred at this time. Deposition of unit IVa was under way before 14 000 BP and had ended well before 9000 BP when alluviation of sub-unit IVb reached a peak. Sub-unit IVb consists of flood silts and sands along the floodplains of major streams, and suggests a return to seasonal overbank deposition with stable soil mantles on valley margins. Such conditions, with an inferred climate similar to the present, probably began soon after 10 000 BP. Member V of the Riverton Formation consists of two parts and was laid down after 11 m of incision into earlier fills. The older sub-unit Va was deposited as the floodplains of the Vaal and its tributaries underwent extensive low energy aggradation, with local hydromorphic conditions. The fills of small valleys, which are now being dissected by modern gully erosion and their associated colluvium, are sandy but humic and were derived from the weathered aeolian sands of the interfluve areas. A pollen sample from this unit indicates a well developed grass cover. To explain coeval erosion of these soils and a marshy floodplain environment, Butzer (1984*a*) suggests that a moderate increase in rainfall and surface runoff occurred during the mid-Holocene prior to 2700 BP. The younger sub-unit Vb represents moderate energy alluviation in the period 2700–1300 BP.

Deposits at Welgevonden consist of intercalated colluvial and alluvial sediments of silt to loam and clay loam textures (Butzer 1984*a*). They preserve a record of moderate to high energy Holocene alluviation, which began some 5000 BP. Lower energy alluviation, with deposition of fine grained sediments in a hydromorphic and well vegetated riparian environment, continued from 4000 BP to the nineteenth century when it was terminated by land use changes.

Pans

The western Orange Free State and adjacent areas of the northern Cape Province contain one of the major concentrations of pans in Southern Africa (Wellington 1945; Goudie and Thomas 1985). The pans were formed by deflation of surficial sediments in dry seasons, and periods following weathering of underlying Ecca and Dwyka shales in wet seasons and/or periods. In some areas, the record of periods of deflation is contained in small lunette dunes on southern and south-eastern pan margins; elsewhere, deflated aeolian sediments blanket the terrain. Within the pan basins, wave cut notches, and lacustrine and spring deposits record periods of wetter climates with increased spring discharge and high water levels in pans. Palaeosols indicate periods of relative geomorphic stability.

The Alexandersfontein basin east of Kimberley is one of the largest in the region. Its record is complemented by those from Voigtspost, Deelpan, and Klipfontein (Butzer 1984*a,b*). Early in the Late Pleistocene deposits associated with Middle Stone Age artefacts indicate high lake levels, followed by drier conditions and calcrete formation (Butzer 1984*a*, p. 42). Three further major Late Pleistocene and Holocene moist intervals, interrupted and separated by drier spells, can be identified. From *c.* 16 100 BP, or earlier, to *c.* 13 600 BP the Alexandersfontein basin was occupied by a +12–16.5 m deep lake with an area of 24–44 km^2. At the same time springs were active and hydromorphic soils developed in the floodplains of influent streams (Butzer 1984*a*). Butzer *et al.* (1973) estimated that, even with a 6°C lowering of temperatures, such a lake would require an annual rainfall of 670–860 mm to sustain it, compared with some 400 mm today. Spring activity ceased in a dry interval around 14 500 BP and after *c.* 13 600 BP the springs dried out entirely. The period from *c.* 13 500 to *c.* 11 500 BP was one of fluctuating wetter and drier climates, comparable with that observed from the pollen record at Aliwal North by Coetzee (1967). Pollen spectra from deposits laid down in the dry phases of this period are composed of more than 90 per cent Aizoaceae and are similar to modern Karoo-type vegetation (Butzer 1984*a*). There was a short period of renewed spring activity 12 900–12 300 BP after which erosion of lacustrine and spring deposits occurred. Around 11 500 BP spring activity was renewed. Lakes formed in the basins of Alexandersfontein, Voigtspost, Klipfontein (Bushman's Fountain) and Deelpan in the period around 7700–6300 BP. Pollen spectra from Voigtspost suggest that bush (including *Combretum* and *Boscia* spp.) and possibly thornveld vegetation occurred around the pans, indicating warmer and wetter conditions prior to 6000 BP (Horowitz *et al.* 1978; Scott 1984a). A period when conditions were at least as dry as today occurred between *c.* 6000 and 4500 BP, marked by a break in deposition and soil stability. Archaeological visibility in the region was low during this period (Butzer 1984*a*). Renewed spring activity and shallow lakes in pans are evident for the period 4500–1300 BP with short dry intervals around 3100 and 1800 BP. Organic soils formed in locally wet habitats. Pollen contained in hyaena coprolites from Deelpan is dominated by Gramineae (90 per cent) indicating more grassy vegetation, whilst other areas record little change (Scott and Klein 1981). After 1300 BP there occurred a further dry period at Voigtspost which lasted until around 400 BP when conditions were drier than today (Horowitz *et al.* 1978) with deflation of pans *c.* 1200–800 BP.

5.9 Chronological summary

During the period 40 000 to around 12 000 BP, evidence from frost weathered debris in caves, the isotopic composition of animal bones, and latterly pollen, all indicate that the region experienced generally lowered temperatures. There is some evidence from pollen and lake levels that conditions were moister, especially from 16 000 to 12 000 BP. The Pleistocene–Holocene transition appears to have been a period of alternating wetter and drier climates, with warming evident especially in the dry phases. The Early Holocene generally appears to have been somewhat drier than at present. A period of maximum temperatures and increased rainfall some 8000–6000 BP is suggested by pollen and high water levels in pans, with drier conditions thereafter, and short spells of increased rainfall and possibly cooler conditions in the Late Holocene.

6 Winter and all-year rainfall regions: the Cape Ecozone

The Cape Ecozone (Fig. 6.1) comprises the southernmost part of the subcontinent, and includes the mountains and valleys of the Cape Fold Belt together with the adjacent coastal lowlands. It is convenient to recognize two subdivisions that are relevant to Late Quaternary palaeoclimates. The western portion consists of a wide, mostly sandy coastal lowland between the Atlantic coast and the mountains. In this region the mountains form a prominent arcuate escarpment extending from Cape Hangklip to Elands Bay. On the coastal plain deeply weathered rocks of the Malmesbury Group and the Cape Granite Suite are overlain by a thick sequence of Tertiary and Pleistocene marine and terrestrial sediments (Rogers 1982). The western Cape has a mediterranean-type climate with mild wet winters and hot dry summers, and an annual rainfall that ranges from 500–600 mm near Cape Town to 200–280 mm to the north at Lamberts Bay. Orographic influences strongly affect

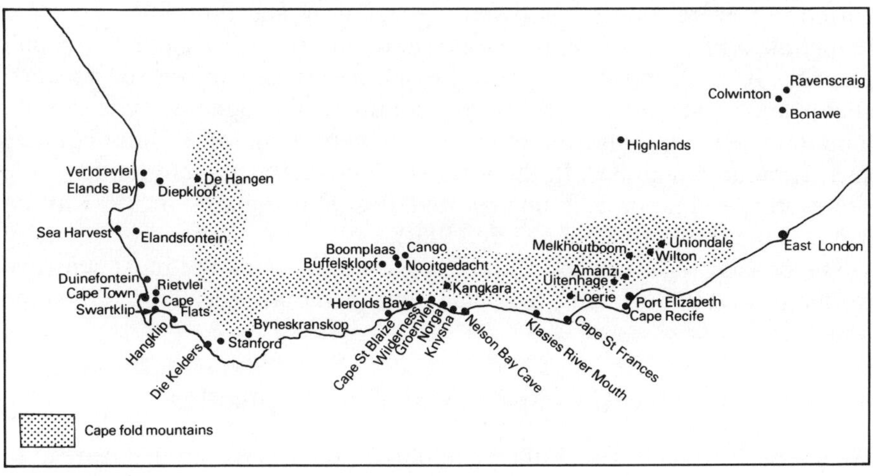

Fig. 6.1 Location of sites in the Cape Ecozone.

precipitation close to the mountains and up to 1500–2000 mm occurs on west and north-west facing slopes in places.

The southern Cape coast consists of a series of eastward-facing sigmoidal bays formed by resistant outcrops of Cape Supergroup quartzites. The coastal platform, at an altitude of about 200 m, narrows from 20 km or more in the west to 2–5 km in the east. It is deeply dissected by rivers draining the coastal mountains, the Outeniqua and Tsitsikamma ranges, which rise to an altitude of 1600 m. Inland, Bokkeveld shales form the substrate in the intermontane valleys such as the Langkloof and Little Karoo. The valleys lie between east-west trending mountain ranges, the highest of which is the Swartberg rising to 2100 m. The climate of the southern Cape is warm temperate with rainfall in all seasons and maxima in spring and autumn. Topography again influences the amounts received which range from 500–600 mm on the coast to 1000–1200 mm in the mountains. The valleys of the Little Karoo lie in a rain shadow and receive only 200–400 mm per annum.

Fynbos vegetation (Proteaceae, Ericaceae, and Restionaceae) predominates in the winter rainfall belt, both on the mountains and along the coast where soils are low in nutrients. Patches of Afromontane forest occur where rainfall is highest and tends to be biseasonal along the southern Cape coast. Renosterveld, where the vegetation is dominated by Compositae, mainly the renosterbos *Elytropappus rhinocerotis*, flourished in pre-agricultural times on plains and intermontane valleys where soils are derived from shale substrates.

In historic times the larger mammal fauna was dominated by browsing ungulates such as bushpig, bushbuck, blue duiker, grey duiker and Cape grysbok. One larger grazer, the Cape buffalo, was fairly common. Cape hartebeest were numerous where vegetation was suitable, as were bontebok and the now extinct blue antelope. Roan occurred in small numbers in the forested areas to the south-east and the mixed feeders, eland and steenbok, were fairly common throughout. Vaalribbok, mountain reedbuck and mountain zebra were common in hilly country and on mountain slopes. In the western Cape elephant were found in the well-watered areas and a small population still remains in the Knysna forest in the southern Cape (Klein 1980).

The coastal region in the Cape Ecozone is perhaps the most sensitive to change and the evidence for palaeoclimatic conditions there is of some importance.

6.1 Dune reactivation and stabilization episodes

Weathered aeolianites and unconsolidated dune sands are widespread in the coastal zone of the Cape region. They occur east of Algoa Bay

(Ruddock 1968; Illenberger 1986) where the Alexandria dune field is the largest in the province, and in the Humansdorp, Tsitsikamma, Knysna, and Wilderness areas (Martin 1962; Butzer and Helgren 1972). They are more than 150 m thick in the Knysna forest and are up to 300 m thick at Wilderness.

Martin (1968) recognized four successive dune belts in the Wilderness area. Two further submerged dune ridges were identified offshore of the Wilderness by Birch *et al.* (1978). Butzer and Helgren distinguished at least six generations of aeolianites and blown sands on the basis of palaeosol stratigraphy.

As Butzer and Helgren (1972) point out, the formation of coastal dunes is a function of close proximity to beaches and a plentiful supply of sand. They are thus not primary evidence for climatic changes although the existence of multiple coastal dune ridges reflects changing Pleistocene sea levels. Martin (1962) identified dune ridges I, II and III and the oldest parts of ridge IIIa around Wilderness as older than the Last Interglacial marine transgression to +5–12 m. This transgression is marked by littoral deposits containing thermophile mollusca which are correlated with the Swartkops fauna of Last Interglacial age (Davies 1980). The submerged ridges recognized by Birch *et al.* (1978) off the southern Cape coast relate to sea levels 35 and 45 m lower than at present and are thought by them to represent stages of the regression of sea levels during the Last Glacial.

Renewed accumulation of dunes equivalent to ridge IV of Martin (1962) took place at some time after 16 000 BP between Wilderness and Plettenberg Bay (Butzer and Helgren 1972). This was probably in response to a massive input of sediment as waves reworked fluvial and aeolian deposits laid down on the exposed shelf during the −130 m sea levels of the Last Glacial Maximum (Dingle and Rogers 1972*a,b*). These dunes were therefore formed during the southern Cape equivalent of the Flandrian transgression.

Dating evidence comes from the Robberg Peninsula near Plettenberg Bay where transgressive aeolianites accumulated in the terminal Pleistocene and Early Holocene. Profiles were mapped by Butzer and Helgren (1972) on the 'Island' joined to the southern shore of the peninsula by a tombolo. Inorganic carbonates in the A1-horizon with a radiocarbon date of 16 000±220 BP (UW-198) give a terminus ante quem for the development of Palaeosol 1. Aeolian sands overlie Palaeosol 1, and Palaeosol 2 developed in the upper part of these. Shell of the land snail *Achatina zebra* near the base of an even later layer of aeolian sand overlying Palaeosol 2 has been radiocarbon dated to 7030±120 BP (UW-201). This most recent phase of mobilization of sands thus began about 7000 BP (Butzer and Helgren 1972).

Some indications of palaeowind directions are reported by Butzer and Helgren (1972) who inferred that deposition of dunes on the island off the Robberg Peninsula during the Pleistocene took place when strong winds were dominantly from the south-east and south-west. Modern strong winds in this area are westerly to north-westerly, suggesting changes in storm winds during this period.

Further west, strong winds during the Late Pleistocene are assumed by Rogers (1986) to have led to the formation of the thick calcrete-capped aeolian deposits at Rooikrans east of Stillbay. This Langebaan Limestone Member is thought to date to the maximum regression of the Last Glacial.

Soil development on aeolian sands implies periods of stability which may result from reduced sediment supply and/or increased vegetation cover. The degree and type of pedogenesis may reflect specific climatic conditions. Butzer and Helgren (1972), Helgren and Butzer (1977), and Butzer (1984*b*) recognize three generations of Holocene palaeosols which formed during the periods *c.* 8000–7000, 4850–1300, and 500 BP in the southern Cape. The Beacon Island soil is described as a weakly developed humic soil with signs of incipient podsolization that formed in the period prior to the radiocarbon dated horizon on the Robberg 'Island' of 7030±120 BP (Butzer and Helgren 1972). They see it as coincident with forest expansion and peat formation in the Wilderness area reported by Martin (1968). Butzer and Helgren suggest renewed stabilization occurred from *c.* 4850–1300 BP coeval with further peat formation and a rise in sea level to 2.5 m. The most recent period of stabilization is said to date to *c.* 500 BP (Butzer 1984*b*) and all three are interpreted as having formed in periods when dunes were stabilized by increased vegetation cover resulting from higher precipitation. Earlier periods, i.e. the Early Holocene, as well as the mid-Holocene are thought to have been relatively dry.

Butzer and Helgren (1972) also recognize an earlier Late Pleistocene weakly developed red yellow podsolic soil which they have named the Brenton Soil. It is widely developed on aeolianites, colluvial deposits, and shorelines of Last Interglacial and early Last Glacial age, and represents the cumulative effects of decalcification, humification, and iron mobilization. It appears to differ only in degree of development from the Holocene humic soils and has again been interpreted as evidence for increased precipitation.

In the western Cape, Butzer (1984*b*) suggests that the process of dune accumulation was not continuous, and was interrupted by a series of periods of stabilization and soil formation, probably in moist conditions. At Van Riebeeckstrand, the dunes resulting from the regression of the Last Glacial are interrupted by three palaeosol horizons which document, according to Butzer (1984*b*), three periods of subhumid climates between *c.* 25 000 and 15 000 BP, possibly followed by further soil development prior to Holocene calcrete formation. At Swartklip and Stanford, single palaeosol horizons are recognized by Butzer (1984*b*) as part of this regressional dune sequence although only one radiocarbon date is available, on bone from Melkbosstrand at *c.* 22 000 BP. The basis for assuming that these coastal palaeosols indicate wetter conditions is that they imply that dune sands were vegetated. However, at times of lower sea level they could have become vegetated as the shoreline regressed and may not necessarily indicate higher rainfall.

As part of the same study, Butzer (1984*b*) has interpreted the formation of calcretes in the Late Pleistocene regressional dunes as an indication of drier climates in the Early Holocene. More humid conditions in the mid-Holocene are suggested by the formation of weakly humic soils at Van Riebeeckstrand, Swartklip and Stanford between two phases of Holocene sand accumulation (Butzer 1984*b*).

6.2 Evidence for former sea levels in the western and southern Cape

Tankard (1976*b*) has made an extensive study of the deposits and shorelines of the south-western Cape, an area which he believes was tectonically unstable throughout much of the Late Cainozoic, with tilting occurring about an axis running north-west from Cape Agulhas through the Orange River mouth. However, Pether (1986) summarizes evidence for the Early Quaternary '30 m Package' on the west coast and concludes that it is only through securely correlated sedimentary successions and an integrated study of sedimentary loading history and crustal response that sea-levels can be correlated around the Southern African coast.

6.2.1 Last Interglacial sea levels

It has long been assumed that the Last Interglacial sea level was between 6 and 8 m above the present along much of the western and southern Cape and Natal coast. Tankard (1976*b*) has described shell beds on the seaward edge of a granite pluton at Vredenburg and near the mouth of the Berg River, as well as extensive barrier beach and breaker bar complexes of the St. Helena Bay coast, as related to a major Last Interglacial stillstand at +6.3 m (Tankard 1976*b*). Shell deposits attached to rocks on the southern side of the Verlorenvlei estuary are at a maximum elevation of +5 m. Molluscs still attached to rocks in the living position indicate a sea level of +6 m and a correlation with the major stillstand at +6.3 m. In Saldanha Bay and Langebaan Lagoon, Birch

(1976) recognizes a transgression to 7–10 m which may be correlated with that recognized by Tankard (1976*b*). Tankard (1976*b*) suggests further evidence for a slightly younger transgression to +2.5–3.5 m in the form of shell beds banked against those of the 6.3 m transgression and lagoonal, and tidal flat sediments at an altitude of 2–2.5 m. These shorelines, together with another at about modern sea level indicated by beach rock, are regarded by Tankard (1976*b*) as being of Last Interglacial age.

The molluscan species associated with these shorelines consist of two groups (Tankard 1975). These are cool water open coast species, and warm water estuarine and lagoonal species. The open coast is characterized by species living today in similar environments on the west coast, but the group from estuarine and lagoonal facies contains many more species (more than 30 per cent) that are tropical and thermophilous. Some of these species are found today as much as 2000 km away in West Africa. The presence of warm water mollusca indicates that minimum water temperatures in west coast embayments were 4–6°C warmer than at present during the Last Interglacial and did not fall below 18°C. Tankard (1975), and Tankard and Rogers (1978) suggest that during the +6–7 m transgression, the west coast became extensively embayed. Estuaries, such as those of the Verloren and Berg rivers, became shallow bays, with marine conditions extending as much as 15 km inland (Tankard 1976*b*). Lagoons were created along the low-lying sandy coast. In these environments, strong solar heating raised water temperatures to levels where thermophile mollusca could survive and reproduce, even though the open coast water temperatures were not significantly different from those of today.

The palaeoclimatic implications of these mollusc assemblages are important. Southward migration of thermophile mollusca on the west coast was aided, according to Tankard (1975), by a narrower gap between the warmer Angolan and cold Benguela currents, and possibly by a strengthened coastal current counter to the northerly flowing Benguela. These phenomena were caused, in Tankard's view, by a more southerly position of the South Atlantic anticyclone, and possibly weaker circulation in the South Atlantic, leading to less upwelling of cold bottom waters. Strong solar heating is also indicated and would be promoted by thermal maximum conditions of an interglacial period.

Further evidence for warm and dry conditions on the west coast during the Last Interglacial comes from the evaporite deposits along the coast south of Cape Deseada near Elands Bay (Tankard 1976*b*). The evaporites contain over 80 per cent gypsum and consist of a series of rhythmically laminated calcium-gypsum-halite units each 2 mm thick. They overlie shelly calcareous sands correlated with the 6–7 m Last

Interglacial transgression and are in turn covered by dune sands. According to Tankard (1976*b*), the evaporites were produced by flooding of lagoons behind a beach bar, followed by evaporation in which calcium carbonate was precipitated first, followed by gypsum and then halite. The species composition of ostracods suggests that, from time to time, freshwater from landward run-off flooded the pans.

The existence of these evaporite deposits along the west coast is regarded by Tankard (1976*b*) as an indicator of a hot dry climate in the area with minimal surface run-off and high evaporation rates that most likely dates to the Last Interglacial.

On the southern Cape coast, Rogers (1905) was the first to show that Middle Stone Age assemblages consistently overlaid the 6–8 m beach. This has been confirmed at Klasies River Mouth where there are eight caves in the complex. Five of them (1A, 2, 3, 4, and 5) were cut by sea levels up to 18 m. The other three (1, 1B, 1C) were cut by more than one high sea level up to 8 m (Deacon *et al.* 1986). After excavations in the Cave 1, 1A and 1B complex, Butzer (1978), and Singer and Wymer (1982) described the base of the Middle Stone Age sequence in Cave 1 at 8 m above present sea level. They interpreted it as post-dating the 6–8 m Last Interglacial transgression and therefore dating to the end of oxygen isotope stage 5e because people were able to occupy the cave only once the coastline began to retreat. More recent investigation of the site by Deacon *et al.* (1986) has shown that beach material underlies a calcareous crust with coarse angular clasts and few artefacts. The Middle Stone Age (MSA) I units overlie the crust and speleothems have grown on top of these oldest occupation deposits (Fig. 6.2). J. C. Vogel (personal communication) has obtained two ionium dates of 98 000 and 110 000 BP for the speleothem material.

Countering the interpretations of Tankard, Butzer, and Singer and Wymer, Hendey and Cooke (1985), and Hendey and Volman (1986) have argued that the +6–8 m sea level dates to the Early Pleistocene, and that the Last Interglacial transgression was no higher than +4 m. At Klasies River Mouth, Hendey and Volman (1986) argue that although the MSA I occupation was contemporary with oxygen isotope stage 5e [as confirmed by the oxygen isotope content of associated marine shell analysed by Shackleton (1982)], the Last Interglacial sea level was no higher than +4 m at the site. This interpretation would certainly help to explain how people could have occupied the cave at the same time as the marine transgression. However, at Herolds Bay Cave to the west, Brink and Deacon (1982) report that MSA artefacts with shell midden debris directly overlie beach sediments at +5 m, and are in turn overlain by aeolian lenses indicating a regression, coarse angular material that includes bones collected by hyaenas and/or porcupines, and flowstone

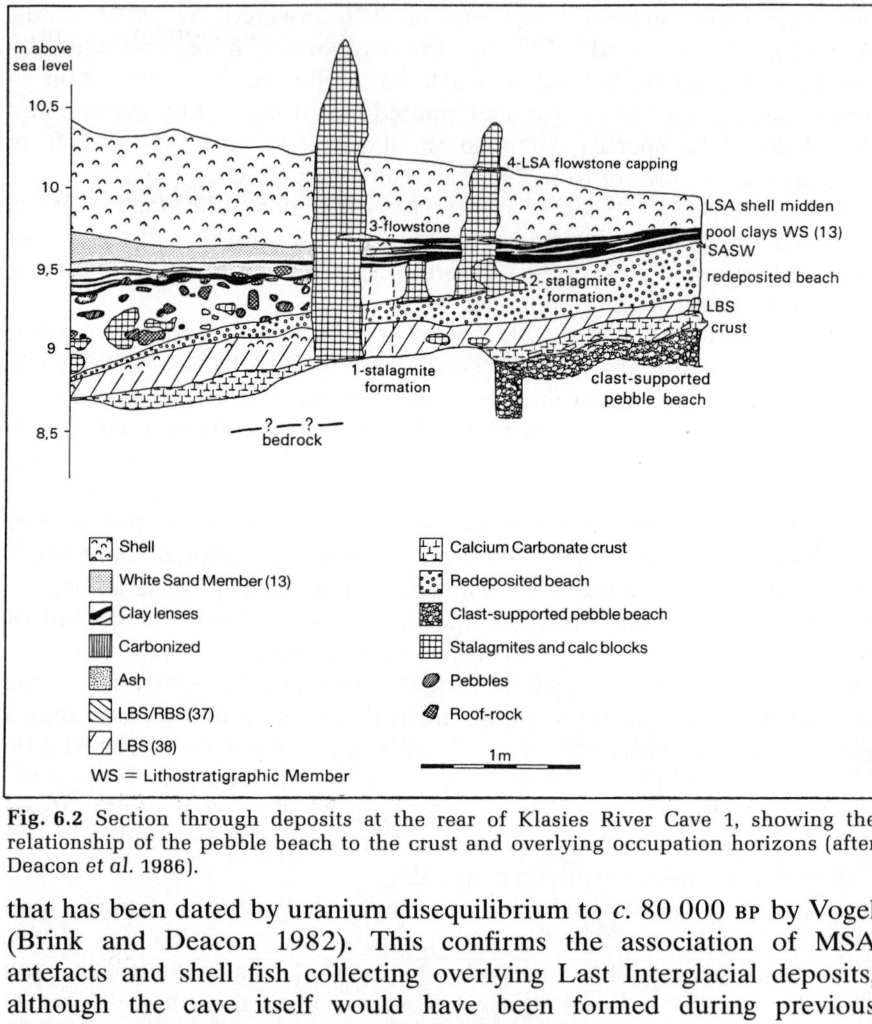

Fig. 6.2 Section through deposits at the rear of Klasies River Cave 1, showing the relationship of the pebble beach to the crust and overlying occupation horizons (after Deacon *et al.* 1986).

that has been dated by uranium disequilibrium to *c.* 80 000 BP by Vogel (Brink and Deacon 1982). This confirms the association of MSA artefacts and shell fish collecting overlying Last Interglacial deposits, although the cave itself would have been formed during previous interglacials (see Hendey and Volman 1986, p. 196).

On the west coast the earlier dating for the 6–8 m beach was suggested in southern Namibia by Corvinus (1983) who described Earlier Stone Age Acheulian artefacts associated with the 'C' beach at this height. A more precise date was offered by Hendey and Cooke (1985) who identified an extinct bushpig *Kolpochoerus paiceae* in a consolidated coastal dune at Skurwerug near Saldanha adjacent to the 6–8 m shoreline. This species is more primitive than the Middle Pleistocene form at the nearby Elandsfontein fossil site, and most closely resembles a similar species dated between 1 and 2 Ma in East Africa. Hendey and Cooke (1985), and Hendey and Volman (1986) point out that middle Pleistocene fossils and Acheulian artefacts are also found post-dating the

6–8 m shoreline at Duinefontein and Bok Bay between Cape Town and Saldanha. They recognize only one Last Interglacial shoreline at about 4 m above present sea level in the Saldanha region.

Hendey and Volman (1986, p. 196) find further support for their dating of the +4 m shoreline at Die Kelders where Middle Stone Age occupation deposits built up over the base of the wave cut cave. Tankard and Schweitzer (1974) described the base of the sequence at 2 m above present sea level, or possibly equal to the present (Tankard 1976*a*). Hendey and Volman suggest that the beach deposits exposed during excavation were emplaced there by the +4 m sea level during oxygen isotope stage 5e, and that formation of the cave began during the 6–8 m sea level of the Early Pleistocene.

At Swartklip, Barwis and Tankard (1983) have recognized four depositional facies between 5.1 and 8 m above present mean sea level. Seaward-dipping beachrock occurs intermittently in the modern intertidal zone and is interpreted as a beach foreshore wedge that originated in warmer climates of the Last Interglacial high sea level. The calcrete capping is thought to have developed in the subsequent lowstand. A later rise in sea level at *c.* 120 000 BP is reflected in a 2-m thick estuarine fill that includes shells of *Tomichia ventricosa*, a gastropod that lives today in aquatic environments ranging from perennially fresh fluvial settings to coastal playa lakes that may be fresh or hypersaline. The third facies, landward-dipping sheet sands which overlie the estuarine fill, is attributed to washover fan deposition. Deposition of the last facies, calcite-cemented sandstone that includes pulmonate gastropods, is believed to date to the start of the Last Glacial as sea level dropped and atmospheric circulation was more intense. The vector mean of these aeolianites is north-west indicating a predominantly south to south-easterly summer wind pattern. Fossil bones in small cavern infills in the beachrock derive from carnivore kills and indicate cooler temperatures than those of the present (Klein 1975, 1980). Klein has suggested the fossils date to a cool substage of isotope stage 5, but Barwis and Tankard (1983) favour a more recent dating between 75 000–80 000 and 40 000 BP., i.e. in oxygen isotope stage 4.

Away from raised beaches, major decalcification and fissure corrosion occurred in transgressional dunes attributed to oxygen isotope stage 5e at Swartklip before the regression of stage 5c. A similar situation obtains at Stanford where major decalcification took place in stage 5d or 5c, suggesting widespread moist conditions at this period (Butzer 1984*b*).

6.2.2 Last Glacial sea levels

Mabbutt (1957) described ferricretes in coastal dune areas near Bok Bay which he attributed to pluvial conditions following regression from the

+6–8 m sea level. Apparently similar deposits were identified in the Berg River valley by Tankard (1976*b*) in dunes formed during the Last Interglacial transgression at Duinefontein (Butzer 1984*b*) and, associated with a truncated podsol, at the Elandsfontein fossil site (Deacon 1964), and would appear to represent periods of increased flooding and high water tables caused by higher rainfall. At Elandsfontein, the ferruginized horizon overlies deposits containing Middle Pleistocene fossils and contains Middle Stone Age artefacts, suggesting an early Late Pleistocene age, possibly coeval with the formation of iron rich horizons in many southern Cape cave sequences.

During the Last Glacial Maximum, sea levels fell to −110 m off the west coast (Tankard and Rogers 1978; Vogel and Visser 1981) or to as much as −130 m on the southern coast (Dingle and Rogers 1972*a,b*; Fig. 6.3) exposing the wide shelf zone and initiating extensive incision of river valleys. North of the major rivers on the west coast, plumes of parabolic and barchanoid dunes were formed by strong southerly winds from an increased supply of sediment derived from the incising rivers and possibly augmented by higher run-off from their headwaters.

Offshore, Flemming (1976) has suggested that the submerged wave-cut platform at Rocky Bank between −30 and −40 m below sea level south-

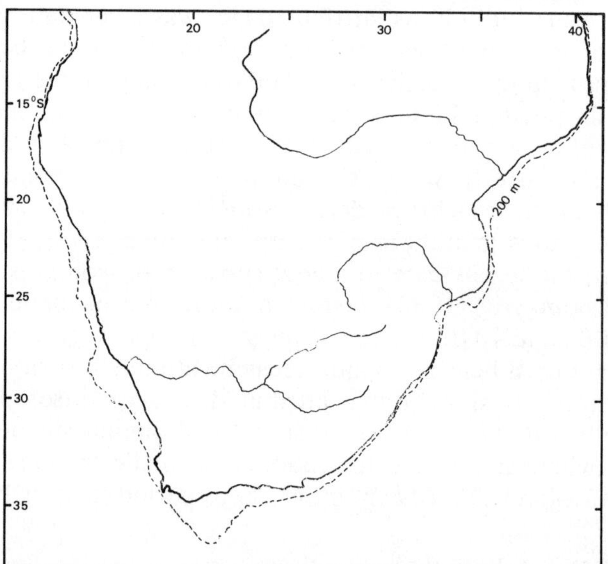

Fig. 6.3 Extent of the continental shelf (200 m below present sea level) indicating the approximate position of the coastline at the Last Glacial Maximum.

east of Cape Point, indicates a sea level 20–25 m below the present, probably during a Last Glacial stillstand. It corresponds to the −20 m submerged cliff along parts of the west coast of the Cape and Namibia. Tankard (1975) suggests a date of between 47 000 and 25 000 BP.

Offshore at Cape St Francis, littoral shells and calcareous algae from depths up to −130 m have been dated by radiocarbon to the Last Glacial Maximum and the subsequent transgression (Dingle and Rogers 1972*a,b*). The oldest is a *Pecten* valve dated to 16 990±100 BP (Pta-182; see Table 6.1). Assuming an even rise in sea level (which probably was not the case), sea level from the onset of the change rose by 1.3 m per 100 years.

On the west coast, Vogel and Visser (1981) have dated a series of littoral shells on the continental shelf between the Orange River and Luderitz. The deepest is at about −90 m dated to *c.* 16 000 BP (Fig. 6.4).

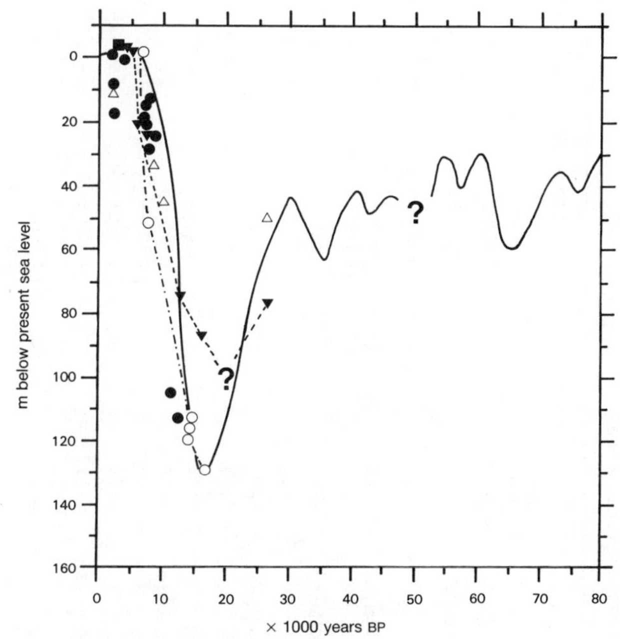

Fig. 6.4 Changes in sea level off the West and Southern African coast from radiocarbon dated shells on the continental shelf off Cape St Francis (open circles), the Zaire coast (closed circles), between the Orange River Mouth and Luderitz (closed triangles), and Mauretania (open triangles) plotted against the sea-level curve from the Huon Peninsula, New Guinea, recalculated after detailed correlation with the oxygen isotope record from core V19–30 by Chappel and Shackleton (1986). The closed square is the +2.8 m date from Elands Bay (Table 6.1). Dates not listed in Table 6.1 are from Giresse and Kouyoumontzakis (1974) and Einsele *et al.* (1977).

Table 6.1. Radiocarbon dates for sea levels lower and higher than present on the west and south coasts of Southern Africa.

Site	Height above or below present (m)	Radiocarbon date (years BP)	Lab. number	Reference
Luderitz	+1.5	1580 ± 50	Pta-417	Vogel and Visser 1981
Saldanha Bay	+1.5–2	2040 ± 50	Pta-1601	Flemming 1977
		2070 ± 60	Pta-461	
		3500 ± 60	Not given	
Groenvlei	+1.5	between 1905 ± 60	Y-467	Martin 1968
		and 6870 ± 160	Y-466	
Elands Bay	+2.8	3820 ± 50	Pta-4041	Yates *et al.* 1986
Luderitz	+3.6	4940 ± 60	Pta-419	Vogel and Visser 1981
Orange River	±0	5750 ± 60	Pta-1351	Vogel and Visser 1981
	−20.4	7130	Pta-1099, 1098	
Cape St Francis	−51	7580 ± 70	Pta-183	Vogel and Marais 1971
Orange River	−75.5	13 300	Pta-955, 951, 957, 959	Vogel and Visser 1981
	−87.2	15 700 ± 160	Pta-1105	Vogel and Visser 1971
	−78.4	27 400 ± 440	Pta-1104	Vogel and Visser 1971
Cape St Francis	−112	14 510 ± 120	Pta-265	Vogel and Marais 1971
	−115	13 670 ± 120	Pta-264	Vogel and Marais 1971
	−120	12 990 ± 100	Pta-185	Vogel and Marais 1971
	−130	16 990 ± 100	Pta-182	Vogel and Marais 1971

6.2.3 Holocene sea levels

Evidence for Holocene sea levels higher than today in the south-western Cape is controversial. Tankard (1976*b*) could find no unequivocal evidence of higher Holocene sea levels. However, Birch (1976) and Flemming (1977) report a +2–3 m sea level from Langebaan Lagoon dated between 6500 and 5000 BP. Fossil oyster beds dated between 6400 and 1800 BP are also reported by Flemming (1977) from the Langebaan Lagoon. They appear to indicate rather warmer waters at this time and may provide a Holocene equivalent of the Last Interglacial warm estuary phase. More recently, Yates *et al.* (1986) have reported a buried cobble beach with shell about 200 m south of the mouth of Verlorenvlei. It has been surveyed at 2.8 m above present sea level and a sample of shell from immediately beneath the cobble layer was dated at 3820±50 BP (Pta-4041; Fig. 6.4). The date has been corrected for shell.

An unusually well developed raised gravel beach extends westwards for about 12 km from the mouth of the Gouritz River to Ystervarkpunt. It is up to 6.4 m above sea level and is interpreted by Rogers (1986, p. 7) as a storm beach of a 2–3 m Holocene sea level high that he equates with that recognized around Langebaan Lagoon on the west coast dating to between 3000 and 2000 BP.

At Groenvlei near Wilderness, there is equivocal evidence for a mid-Late Holocene high sea level. Martin (1968) has identified a transgression from diatoms at +1.5 m that is dated between 6870 ± 160 BP (Y-466) and 1905 ± 60 BP (Y-467).

Further east, Illenberger (1986) has studied dune field morphology in the Alexandria dune field and explains the existing morphology as the result of pulses in the sand supply to the dune field. The pulsing sand supply could result from small variations in sea level during the past 6500 years, climatic changes, destruction of dune vegetation by pre-historic human activity, or bushfires. If sea level changes are responsible, then the full sequence of sea level changes should be represented. Sand would have become mobilized only at times of present-day sea level or higher. With sea level regression, the Sundays River would have cut into the sands on its way to the coast and dune field development would have been halted. The deposits are undated, but Illenberger infers that vegetated dune fields inland of the mobile ones probably date to the Last Interglacial and older. By mapping the morphology of the dune fields in relation to the coast and the sand supply, he has suggested the following sequence of sea level changes in the mid-Late Holocene: sea level at +3 m at 6500 BP, sea level at −15 m 5000 BP, sea level close to the present 3500 BP, sea level at −5 m 2500 BP and sea level the same as at present from 500 years ago (Illenberger 1986, p 30 ff). This is a much more complex series of sea level changes than has been recognized elsewhere along the Southern African coast, and needs to be dated and tested at other localities.

6.3 Alluvial deposits of the southern and western Cape

The fluvial history of the major rivers is linked closely to the history of Pleistocene sea level fluctuations with aggradation during transgressions and high stillstands, and incision during regressions. The major regression to between −110 and −130 m during the Last Glacial Maximum destroyed much of the evidence for former aggradational phases.

The major rivers of the southern and south-eastern Cape follow east-west structural trends. Smaller streams flow south from the coastal mountain ranges and are deeply incised into the coastal platform. River terraces are infrequently preserved, probably as a result of massive over-deepening of valleys during lowered glacial sea levels. The terrace deposits which survive seem mainly related to sea level fluctuations, although their character does give some indication of the climatic conditions existing when they were laid down. In the Bietou River valley, for example, Butzer and Helgren (1972) interpret strongly sorted, torrent-bedded coarse gravels derived from the Tertiary Keurbooms Formation as evidence for rapid run-off in the Last Interglacial.

Holocene fluvial deposits are relatively well preserved. They consist generally of fine, humic loams. Near Loerie three generations of deposits are recognized by Butzer and Helgren (1972). Terraces at +12 and +5 m are composed of colluvial sandy loams derived from Cretaceous shales and sandstones. The second channel fill is 2–4 m thick and includes sands, humic loams, and peats. The peats are derived from floodplain swamps and have radiocarbon dates of 4010 ± 70 BP at the base and 1330 ± 110 BP at the top. They formed at a time of stabilization, and Butzer and Helgren correlate them with soil formation on dunes near Plettenberg Bay and peat formation in the Wilderness lakes (Martin 1962). A period of incision after 1000 BP was followed by aggradation of coarse sands in the lower parts of stream valleys implying a period of Late Holocene geomorphic instability that was induced possibly by farming (Butzer and Helgren 1972).

6.4 Slope deposits in the southern and western Cape

Butzer and Helgren (1972) note the existence of widespread angular colluvium and slope breccias throughout the southern Cape. On the seaward slopes of the Tsitsikamma and Outeniqua mountains these appear to grade to lower sea levels. They argue that such debris mantles are not consistent with modern slope processes and vegetation covers and conclude that the debris was produced by intensified frost weathering and the colluvia from intensified slope processes operating under a reduced vegetation cover. They are not periglacial deposits as suggested by Linton (1969).

Goede (unpublished) has observed that screes and block streams are absent from the Swartberg in the southern Cape, but are widespread in the Hottentots Holland and other south-western Cape mountains. These features were probably produced by frost wedging of Table Mountain quartzites during the Last Glacial and previous glacials. Moisture as well as cold temperatures are necessary to generate frost weathering and Deacon *et al.* (1984*a*) suggest that the absence of block streams in the Swartberg indicates that the season of maximum cold did not coincide there with the season of precipitation. The coincidence of cold and moisture in the south-western Cape suggests that the rainfall regimes during glacial maxima were different in these two regions.

In the western Cape, Butzer (1979) has interpreted a series of deposits in the Tunnel Section on the seaward end of Bobbejaansberg at Elands Bay (referred to incorrectly as the Elandsberg) although his dating of the sequence differs from that of Tankard (1976*b*). Butzer describes a series of reworked aeolian sands and talus mobilized from the quartzite slopes of the Bobbejaansberg. The talus deposits are seen as reflecting episodes of frost weathering, and large scale mass movement and colluvial deposition, with intervening periods of aeolian accumulation and pedogenesis. He suggests that frost shattering was a major process in the production of the angular debris of units 4 and 5. He considers that unit 4 was deposited in isotope stage 5b and cemented in stage 5a, and that unit 5 can be correlated tentatively with isotope stage 4. If this interpretation is indeed correct, then these slope deposits provide a further example of intense frost weathering in the western Cape which is frost free today.

A period of slope stability, allowing pedogenesis and calcrete formation, followed deposition of unit 3. Butzer (1979) suggests that this marks a period of leaching and mobilization of calcium carbonate from overlying calcareous sands which were laid down under drier conditions. The period of soil formation he suggests implies slightly wetter conditions which he correlates with oxygen isotope stage 5c or 5d and he dates the subsequent dry period of calcrete formation to stage 5a.

6.5 Coastal caves and cave deposits

Deposits in coastal caves at Nelson Bay, Klasies River Mouth, Die Kelders, and Elands Bay have provided palaeoclimatic information from faunal, floral, and sedimentological analyses.

6.5.1 Nelson Bay Cave

Nelson Bay Cave faces south on the Robberg Peninsula at Plettenberg Bay in the southern Cape (Klein 1972*a*; J. Deacon 1978). It was formed at +15–20 m by wave action during periods of high sea level at the contact between Table Mountain series quartzites and overlying Cretaceous sandstones. Butzer has identified basal deposits in the cave as beach sands and gravels of Last Interglacial age. There follows a sequence of fine loamy deposits, with additions of beach derived aeolian sands, which include Middle Stone Age artefacts. Temperate, warm, and moist climates are inferred from the fine grained textures and well weathered debris which dominate lower members of the deposits. From this, Butzer (1984*b*) concludes that most of the period covered by oxygen isotope stage 5 was relatively moist. The uppermost Middle Stone Age deposits (the Black Loams) contain abundant angular quartzite rubble which Butzer (1973*a*) regards as having been derived from frost accelerated roof spalling. This, in terms of the present frost free climate, implies temperatures at least 10°C lower than today at this period. Percolating iron rich waters led to post-depositional ferricrete formation in the topmost loamy horizons. For this event, Butzer (1973*a*) suggests a very wet environment outside the cave, with acidic soil conditions and a seasonally wet cave microenvironment, probably at the time of formation of the podsolic Brenton Soil. A thick rubble horizon follows and was apparently developed in conditions of slow temperate weathering of the cave roof. Late Pleistocene deposits in Nelson Bay Cave include two more ferruginous horizons suggesting further periods of wet climate and a perennially moist cave environment after about 17 000–16 000 BP and, with lesser intensity, in the period between about 12 000 and 10 500 BP (Butzer 1984*a,b*). In both instances the material selected for radiocarbon dating was ostrich eggshell and charcoal from fires made by the cave inhabitants, so the periods of ferruginization are younger than the radiocarbon determinations. Frost generated rubble increases to a secondary maximum around 18 000 BP, suggesting to Butzer (1984*a,b*) a cold environment in the period from about 19 000 to 16 500 BP. Holocene strata are dominated by occupation debris, but record a strong input of aeolian sands at 5500 BP and minimal sedimentation after 4000 BP. This is in line with the periods of dune accumulation and stability recognized from the palaeosol stratigraphy of Butzer and Helgren (1972).

In the large mammal fauna there is a range of species in the deposits dating between 18 000 and 12 000 BP that was unknown in the vicinity during historic times and is not recorded there in prehistoric samples dating within the last 10 000 years. These are all migratory grazers such as wildebeest, bontebok, springbok, quagga, and warthog, all of which are indicative of grassland vegetation (Klein 1972*a,b*, 1980, 1984). They were gradually replaced after 12 000 BP by the non-gregarious browsers typical of the present-day environment, but it is only within the last 5000 years that the historic forms dominate completely (Klein 1983, p. 119). After 12 000 BP, fish, marine mammals, and shellfish remains were brought to the cave as the sea level rose and the coastline approached its present position. Changing sea temperatures from cooler to warmer at about 11 000 BP are possibly reflected in the shift from the black mussel, *Choromytilus meridionalis*, which today occurs in cooler waters of the western Cape, to the brown mussel, *Perna perna*, found in the vicinity of the site at present (Klein 1972*a*). Oxygen isotope readings on shells from the site have been used to show that they were collected during the winter months (Shackleton 1973).

A sample of micromammal bones from units dating between *c.* 18 700 and 16 700 has been analysed by Avery (1982*a*). The community structure indicates the presence of grassland and a cold, dry climate (Avery 1982*a*) and corroborates the larger mammal data from the Last Glacial Maximum.

6.5.2 Klasies River caves

As mentioned above (section 6.2), eight caves have been investigated to varying degrees in the Klasies River complex. The most reliable palaeoenvironmental evidence comes from the long sequence excavated in the 1960s by Wymer (Klein 1976; Butzer 1978; Singer and Wymer 1982) and in the 1980s by Deacon *et al.* (1986) in caves 1, 1a, and 1b. Middle Stone Age occupation horizons have built up over 15 m in the main 1/1A complex. The oldest occupation horizons (MSA I) overlie a beach gravel, and speleothem material above both has ionium dates older than 100 000 BP. The first occupation therefore probably post-dates the peak high sea level of the Last Interglacial. Although dating for the top of the MSA is not possible yet because it is beyond the limit of the radiocarbon method, there is a long occupation hiatus that probably dates between the end of the main Middle Stone Age occupation about 50 000, and about 5000 BP when Later Stone Age people occupied the site. The deposits are unusual in the Cape Ecozone in that they include both well preserved bone and a shell midden that dates to the Last Interglacial and the onset of the Last Glacial. Human skeletal remains from the earliest MSA horizons at the Cave 1/1a/1b complex have been assigned to *Homo sapiens sapiens* (Singer & Wymer 1982) and represent some of the oldest remains in the world of anatomically modern people.

The sediment analyses at Klasies River provide little in the way of secure palaeoclimatic information. Although Van Zinderen Bakker and Butzer (1973) suggested that there was abundant frost generated roof spalling early in the Last Glacial, Butzer (1978) modified this view as a result of additional sediment analyses and suggested that mechanical weathering was unimportant in the development of the cave. Butzer (1984*a,b*) recognizes stalagmite and flowstone development, indicative of external decalcification, in deposits assigned to late in oxygen isotope stage 5e, late 5d to early 5c, and the 5a/4 transition. This appears to provide further evidence of moist conditions in the period when the Brenton Soil was forming, estimated by Butzer (1984*b*) to be during the Last Interglacial. Indications from the sediments of a major cold interval in substage 5b as, for example, the frost shattered debris in Nelson Bay Cave at this time, is corroborated by oxygen isotope ratios in shell which indicate a temperature decline of 10°C at this time (Shackleton 1982). Limited flowstones in Later Stone Age Holocene strata are dated from 4850 to 4500 BP and after 2450 BP and also suggest wetter conditions at these times (Butzer 1984*a,b*).

Relatively large samples of larger mammal remains come from Klasies River caves 1, 1a, 1b, and 1c. The oldest assemblage, MSA I, includes 28 per cent grazers (i.e. equids and alcelaphines) and 72 per cent small browsing antelope which is a similar proportion to that observed in the late Holocene assemblages at the same site (33 and 67 per cent, respectively). The overlying MSA II has even fewer grazers (14 per cent) than browsers (86 per cent). It has been shown that grazer-dominated assemblages are associated with cooler and possibly drier climatic conditions (Klein 1983) so it may be assumed that these two assemblages date to the warmest period of the Last Interglacial (Klein 1976). The combined faunal sample from the overlying Howiesons Poort and MSA III and IV, on the other hand, is dominated by grazers (53 per cent) indicating cooler conditions at this time.

The Shannon index of general diversity in the micromammalian samples from the MSA II and Howiesons Poort shows moderate conditions without extremes of temperature or rainfall, and with low amplitude fluctuations that were intermediate between interglacial and glacial maxima (Avery 1982*a*). Analysis of samples from the more recent excavations by Deacon *et al.* (1986) has confirmed these observations and Avery (1987) concludes that this micromammalian sequence, which includes the MSA III, but does not include samples from the earliest MSA I, dates between oxygen isotope stages 5d and 5a. Thackeray (in press) has calculated from a factor analysis that temperatures were probably lower than at present and became lower through time, but were not as low as during full glacial times.

Shell middens in the lowest MSA levels are amongst the earliest evidence in the world for the exploitation of shellfish by Stone Age people. The tolerances and relative abundances of various species indicate also that ocean temperatures in the lowermost units were warmer than higher up in the midden sequence and the Howiesons Poort and overlying MSA levels do not contain as much shell as the lowest deposits. This is a clear indication of the retreat of the coastline south of the sites (Voigt 1973, 1982; Deacon *et al.* 1986). Oxygen isotope analyses of the shell show that the characteristic oxygen-18 signature of the Last Interglacial ocean is recognizable in the oldest MSA I shell samples (Shackleton 1982) and confirm a gradual cooling trend through the Howiesons Poort to the MSA III that H. J. Deacon (in preparation) correlates with the major regression at the end of oxygen isotope stage 5 dated in the sea level curve of Chappell and Shackleton (1986) from the Huon Peninsula to between 80 000 and 60 000 BP.

Seal bones dating to cooler intervals at Klasies River and to the terminal Pleistocene at Nelson Bay include antarctic and sub-antarctic species, such as the elephant seal, gazelle seal, and crab-eater, but in the warmer Holocene assemblages and in those dating to the Last Interglacial the Cape fur seal that is common along the coast today is often the only species present (Klein 1980, pp. 258–9). This is further evidence for changes in ocean temperatures.

6.5.3 Die Kelders Cave

Palaeoclimatic information from Die Kelders Cave in the south-western Cape (Tankard and Schweitzer 1974; Tankard 1976*a*) largely parallels that from Nelson Bay Cave. Neither site has shell preserved in the Middle Stone Age occupation horizons, but Die Kelders has bone remains.

Die Kelders was formed by wave action on the contact between Table Mountain quartzites and overlying Neogene limestones (Bredasdorp Formation). In turn, these are overlain by aeolianites deposited by southerly winds. As at Nelson Bay Cave, basal deposits are littoral and consist of wave rounded boulders deposited by a Last Interglacial high sea level. Middle Stone Age deposits record an alternating sequence of occupation and non-occupation layers dominated by quartz sands. These are strongly iron stained in lower layers, suggesting a moist cave environment. An important component of the Middle Stone Age deposits is flat, angular flakes of Bredasdorp limestone. They are sharp and free from chemical weathering and are interpreted by Tankard (1976*a*) as roof spall generated by freezing and thawing. They therefore imply much lower temperatures during the period of initial Middle Stone Age occupation and are suggested by Tankard (1976*a*) to date to *c.* 75 000 BP. As with similar deposits at Nelson Bay Cave, they mark the first major drop in temperature of the Last Glacial cycle. The quantity of these roof spalls decreased upwards and they are absent from the uppermost Middle Stone Age deposits.

Middle Stone Age strata are covered by the well weathered boulders of a major roof collapse, thought by Tankard and Schweitzer (1974) to be the result of a catastrophic earthquake. However, the stratigraphic parallels with the rubble horizon at Nelson Bay Cave are clear, so the deposit may represent rather a long period of weathering in damp, temperate conditions. Succeeding deposits are sterile quartzose sands laid down rapidly in a fluctuating standing water body at the back of the cave. They indicate a very wet environment outside the cave, comparable with that inferred by Butzer (1973*a*) for the ferruginous sands, and ferricrete horizons at Nelson Bay Cave and the Brenton Soil.

The top of the sterile sands is marked by a micro-soil horizon which formed, according to Tankard (1976*a*), when the cave was closed, or nearly closed, by dune sands in the Late Pleistocene. This created a stable environment in the cave, isolating it from the temperature extremes required to produce frost spalling and was held by Tankard to be the explanation for the lack of roof spall during the Last Glacial Maximum.

Renewed deposition took place under a different environment and is represented by talus rubble and shelly sands deposited in dry conditions after 6000 BP. The correlation between this period of sand blowing and the mid-Holocene period of dune accumulation identified in the southern Cape by Butzer and Helgren (1972) is evident.

The micromammalian sample at the base of the sequence, assumed to date to the onset of the Last Glacial (Tankard 1976*a*), indicates cooler and wetter conditions than those of the present with extensive grass in the surrounding vegetation. An overlying sample, still within the MSA, shows more moderate conditions that were warmer and drier than previously and with reduced grass in the vegetation mosaic (Avery 1982*a*; Table 6.2). Cooler and possibly wetter conditions prevailed again at the top of the sequence in a unit radiocarbon dated to a minimum age of 31 000 ± 5000–2900 BP on bone apatite (GX-1717) (Tankard and Schweitzer 1974).

Cape molerat samples from the MSA and LSA units at Die Kelders have been measured by Klein (1984), and the MSA individuals are significantly larger. Modern individuals tend to be larger in moister and cooler environments confirming the inferences drawn from the micromammals for cool and wet conditions during part of the sequence.

6.5.4 Elands Bay Cave

Sediments from Elands Bay Cave, at an altitude of 40 m on the western

Table 6.2. Die Kelders palaeoenvironmental inferences based on micromammalian data (after Avery 1982*a*).

Level	Approximate age (years BP)	Vegetation	Climatic inferences
1	*c.* 35 000	Unit 1: extensive grass, dense	?? Wet
2		vegetation on flats, scrub on hills	
3		Unit 2: reduced grass, dense	? General increase
4		vegetation on flats; moderate restioid	in temperature
5		or 'grassy' vegetation on hills	Generally mild
6			conditions perhaps
7			deteriorating
8			gradually
9			
10			
11			
12		Unit 3: slight change in 12; extensive	
13		grass, dense vegetation on flats, restioid or 'grassy' vegetation on hills	
14	*c.* 80 000		

slopes of the Bobbejaansberg, were described by Butzer (1979). At the base of the deposit is frost shattered angular quartzite rubble that is present also on the talus slope below the cave. The basal deposits contain early Middle Stone Age material and may be as old as late Middle Pleistocene. Later sediments include mixed mineral and cultural material that accumulated between 21 000 and 10 800 BP, and shell middens that date to between 10 800 and 7900 BP, and between 3900 and 300 BP.

Butzer's preliminary analyses of the later sediments show that during the period 20 000–11 000 BP there was an increasing input of aeolian sands to the cave that peaked some 12 500 years ago. This is interpreted as reflecting the increased proximity of the shoreline as sea levels rose in the immediate post-glacial period, as well as possibly drier and windier climates (see also Parkington 1986, p. 208). Holocene deposits, according to Butzer (1979), reflect not climatic changes, but the increasing proximity of the shoreline.

Butzer (1979) also examined sediments from Diepkloof Cave some 10 km inland on the southern bank of the Verlorenvlei estuary. He notes a conspicuous frost shattered roof spall in Middle Stone Age (Howiesons Poort) levels, similar to that found elsewhere in the southern and south-western Cape at this period.

At Elands Bay Cave in units dated *c.* 17 000 and 12 000 BP, the molerat samples are very small, but the mean size of individuals suggests conditions were drier rather than wetter. On the other hand, in the early Holocene at the same site, molerats dating between *c.* 11 000 and 9600

BP are larger than those pre-dating them and post-dating 4000 BP suggesting that conditions were somewhat wetter. Unfortunately there are no samples that date to the mid-Holocene in this area, but mid-Holocene molerats at Byneskranskop further south are more similar to the Late Holocene samples there (relatively dry) than they are to the Early Holocene ones (relatively wet) (Klein 1984, 1986). The larger mammal fauna from the Late Pleistocene is typical of other sites in the Cape Ecozone in that grazers predominate and are replaced after 10 000 BP by modern browsing antelope.

Avery (1983) has studied the micromammals from the Holocene levels. The vegetation in the cave surrounds was fairly open shrubland throughout, with some indication of rather wetter conditions around 3000 BP.

From Elands Bay Cave and surrounding sites Parkington (1981, 1986) has correlated the presence/absence and changing composition of mammalian remains, birds, reptiles, shellfish, and fish samples with the rising sea level between about 14 000 BP and the present. In the marine shell samples, limpets dominated the earliest middens dated between *c.* 11 000 and 10 000 BP, and were replaced between 10 000 and 9000 BP by the black mussel that is found in the area today. Parkington (1986) has suggested that this event may signal the re-initiation of cold Benguela upwelling along the west coast after the Last Glacial Maximum and the subsequent transgression. It could also be linked to the configuration of the changing coastline because limpets appear again in some middens dating within the last 1800 years (Parkington 1986, p. 217), but in both cases it signals the onset of modern coastal conditions in this area.

6.6 Caves and cave deposits in the southern Cape interior

6.6.1 Kangkara Cave

Van Zinderen Bakker and Butzer (1973), and Butzer (1984*a*) report periods of increased roof spall and frost debris prior to and during the Last Glacial Maximum at Kangkara Cave inland from Plettenberg Bay. The latter horizon immediately pre-dates *c.* 12 000 BP, but the former is undated and overlies Middle Stone Age assemblages. They report a dating of greater than 44 000 BP, but this is from another site, Paardeberg Cave, some 100 km to the east.

6.6.2 Boomplaas Cave

Boomplaas Cave, in the Cango valley in the foothills of the Swartberg range and about 80 km inland, has a sequence that is fairly complete back to an estimated 80 000 BP. The excavation programme was designed specifically to obtain samples of a variety of palaeoenvironmental indicators so that results from sedimentological and biological data could be compared in the same dated sequence to test both the timing and scale of change (H. J. Deacon 1979).

Although true éboulis sec did not form, angular clasts are more common in units dating between 22 000 and 17 800 BP (H. J. Deacon 1979; Deacon *et al.* 1984*a*). Temperatures were indeed much colder at that time as evident from oxygen-18 measurements from a speleothem in the sealed underground cavern system of the nearby Cango Caves that indicate it was 5.5°C cooler (Vogel 1983).

Pollen samples show a marked contrast between those of the Last Glacial Maximum dated to *c.* 20 000 BP and those dated to within the last 10 000 years (Fig. 6.5). The Last Glacial Maximum samples have a low species diversity with nine pollen types of which five are Compositae

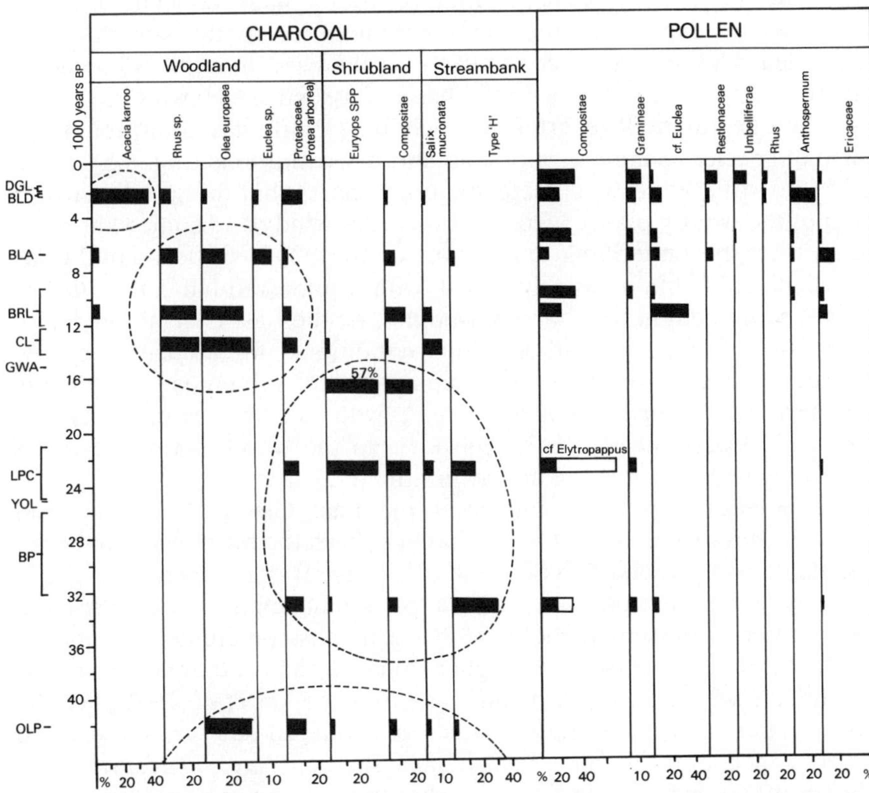

Fig. 6.5 Percentage frequencies of woody taxa (identified from charcoals) and of pollen samples from Boomplaas Cave. Dashed lines circle periods with distinctive plant communities (after Scholtz 1986).

with a single type, cf. *Elytropappus* (renosterbos), accounting for 78 per cent. The Holocene samples, on the other hand, have between 30 and 40 pollen types that are still dominated by Compositae, but with higher percentages of types such as Poaceae (Gramineae), Restionaceae and *Euclea*. There is in addition a small older sample of pollen from the BP Member dated to *c.* 32 000 BP. This is similar to that of the Last Glacial Maximum, but cf. *Elytropappus* accounts for only 58 per cent of the Compositae. The major change between glacial and interglacial vegetation in the surrounds of the cave was therefore in the species diversity with a substantial increase occurring after *c.* 14 000 BP. Whereas the glacial vegetation was dominated by Compositae, the spectra of the Holocene include more trees. There is some suggestion of slightly drier conditions at the end of the Holocene within the last 2000 years with a lower percentage of Ericaceae pollen and fern spores and a gradual increase in *Rhus*-type pollen.

Changes in the species composition of woody plants collected for firewood reflect changes in the plant communities in the surrounds of Boomplaas Cave (Deacon *et al.* 1983). At the base, in units estimated to date to *c.* 70 000–65 000 BP on the basis of associated Howiesons Poort-type Middle Stone Age artefacts, Scholtz (1986) has interpreted the range of species as indicative of cold and dry conditions (Fig. 6.5). From *c.* 55 000–40 000 BP the climate was not as harsh, but the spindly growth habit of the woody plants suggests it was very windy and somewhat more mesic than the underlying unit. From *c.* 40 000–26 000 BP conditions were still cold, but relatively moist with winter rainfall and *Euryops* species prominent in the charcoal samples. At the Last Glacial Maximum *c.* 25 000–17 000 BP, cold and dry conditions are indicated by the charcoal assemblage that is dominated by Compositae (mainly renosterbos, *Elytropappus rhinocerotis*), with low species diversity and no trees (Deacon *et al.* 1983, 1984). From the wood anatomy Scholtz (1986) suggests that rain fell mostly in summer.

The amelioration of climate after the Last Glacial Maximum had clearly begun before 14 200 BP and, judging from the range and number of woodland taxa preserved in the charcoals, there was a marked increase in diversity, although wood anatomy suggests that temperatures were still cooler than at present (Scholtz 1986). Effective precipitation between 14 200 and 12 000 BP was higher than at any other time in the Boomplaas sequence and drought was rare (Deacon *et al.* 1983; Scholtz 1986). Thereafter, in samples dated to about 10 000 BP, there is an increase in composites and thicket taxa such as *Euclea/Diospyros* and *Maytenus/Pterocelastrus* indicating a shift towards warmer and somewhat drier conditions with summer drought stress evident from the wood anatomy (Scholtz 1986).

Thicket taxa became established by about 6400 BP when the range of species indicates warm and very dry conditions with marked signs of drought stress evident from the xylem analysis. Thereafter, *Acacia karroo* was dominant in the woodland in the valley. This shift towards *Acacia karroo* is present also at Buffelskloof rock shelter about 25 km west of Boomplaas (Deacon *et al.* 1983) where it clearly precedes the advent of domestic stock into the valley about 1700 years ago. From the wood anatomy, conditions in the last 2000 years at Boomplaas were warm and mesic with some moisture in the growing season and a low incidence of drought, but higher water stress than in the terminal Pleistocene (Table 6.3).

A sequence of changes can be seen also in the large mammal faunas at Boomplaas and other southern Cape sites. There is a high incidence of

Table 6.3. Boomplaas Cave palaeoenvironmental inferences based on charcoal studies (after Scholtz 1986).

Excavation unit	Approximate age (years BP)	Palaeoclimatic inferences
BLD	*c.* 2000–1000	Decrease in rainfall especially during the warmer months, October to February, results in a contraction of forest vegetation; lower temperatures may also have been experienced
BLA	*c.* 6400	Most xeric conditions during the Holocene; probably little or no rain in the warmer months; long dry, hot summers
BRL	*c.* 10 000–9000	Climate considerably wetter than in the subsequent period, but drier and more seasonal than the preceding period; conditions cooler than during the later Holocene
CL	*c.* 14 000–12 000	Climate is very wet during end-Pleistocene period; total annual precipitation high, rainfall distributed throughout the year and low temperatures contribute towards a consistently favourable precipitation/evaporation ratio; expansion of mesic vegetation types at the expense of other types, but frost incidence or cold winter winds may have been limiting factors
GWA	*c.* 18 000–17 000	Cold and dry throughout the year; scrub vegetation and grassland favoured
LP	*c.* 22 000–20 000	Climate less harsh than subsequently; some rain in summer; karroid scrub vegetation and grassland favoured
BP	*c.* 32 000	When moisture available for growth, temperatures are low; winter rainfall only; cold and dry; sub-Alpine conditions
OLP	*c.* 40 000	Cold and dry, but perhaps without the regular occurrence of harsh climatic extremes; relatively wet
BOL	*c.* 60 000–50 000	Very cold and dry; extremely harsh climate

equids and alcelaphines during the Last Glacial, equids, browsers as well as vaal ribbok and, in some members, *Raphicerus* spp. predominate between 80 000 and 20 000 BP, and a high proportion of browsers is found again in the Holocene. A large number of shorter sequences, as well as a number of carnivore accumulations confirm the general pattern of grazer-dominated larger mammal faunas prior to 12 000 BP and browser-dominated faunas thereafter (Klein 1978*a,b*, 1980) implying a markedly higher grass element in the glacial vegetation than exists today.

The micromammalian community composition at the base of the sequence shows relatively moderate conditions from the mean size of the red musk shrew, *Crocidura flavescens* (Fig. 6.6) and by the Shannon index of diversity, but conditions became increasingly dry and cold in the overlying unit, culminating in the harsh cold and dry environment of the Last Glacial Maximum *c.* 18 000 BP. The means size of *C. flavescens* was greatest at this time and the diversity index was lowest. The implications for vegetation suggest substantially more grass and less scrub before 14 000 BP than after (Avery 1982*a*, 1983; Table 6.4). Thackeray (in press) has calculated a relative temperature index from factor analysis scores for the micromammal sequence (Fig. 6.7) that shows a slightly warmer peak around *c.* 32 000 BP and again confirms that the coldest temperatures were experienced *c.* 18 000 BP. Thereafter, temperatures rose sharply to a peak *c.* 10 000 BP and remained relatively constant through the Holocene (Fig. 6.7). Taking all micromammal palaeoenvironmental indicators together, the most dramatic shift in individual size and

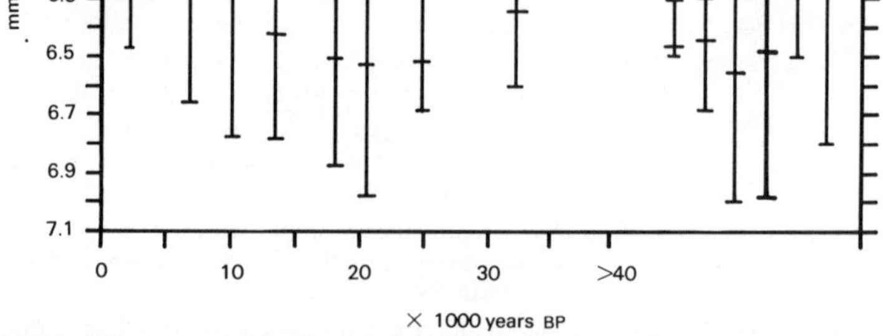

Fig. 6.6 Boomplaas Cave: changes in the range and mean size of individuals in *Crocidura flavescens*, the red musk shrew, as measured by the length of M_{1-3}. The bar on the far left is a modern control sample (after Avery 1982*a*).

Table 6.4. Boomplaas Cave palaeoenvironmental inferences based on micromammal data (after Avery 1982*a*).

Excavation unit	Approximate age (years BP)	Vegetation	Climate
DGL	*c.* 1000–present	Unit 1: scrub on hillsides, sparse ?semi-arid in places	Fluctuating
BLD 3	*c.* 2000	Elsewhere ? dense microphyllous fairly open grass, ?trees on valley	Rapid temperature
BL	*c.* 2500	floor, extensive dense riverside reeds or grass	rise
BLA	*c.* 6400		Fluctuating rapid
BRL	*c.* 10 000–9000		temperature rise
CL	*c.* 14 000–12 000	Unit 2: transitional between Unit 3 and Unit 1	Becoming warmer
GWA	*c.* 18 000–17 000	Unit 3: generally fairly open, ?semi-arid on valley floor; reduced	Coldest
LP	*c.* 22 000–20 000	dense reeds or grass along river; restioid or 'grassy' vegetation on	
YOL	*c.* 25 000	hillsides	Moderate
BP	*c.* 32 000	Conditions more moderate	Mild
OLP	*c.* 40 000	Unit 4: fairly extensive dense vegetation on valley floor; extensive restioid or 'grassy' element on hillsides plus some ?semi-arid vegetation	Cold Relatively warm
BOL	*c.* 50 000	Unit 5: general increase in scrub on hillsides, grass on valley floor and dense riverside vegetation; BOL 5 glacial maximum as in Unit 3	Mild Harsh
OCH LOH	*c.* 60 000	Period of change, ? becoming drier, more open vegetation	Warm, moderate

community composition and diversity occurred between *c.* 16 000 and *c.* 14 000 BP, at least 2000 years earlier than a comparable shift can be discerned in the larger mammal remains.

6.7 Vegetation changes in the Cape Ecozone

As in other ecozones, pollen sequences that extend beyond the Holocene are rare in the Cape Ecozone. The Cape Flats cores, including the one from Rietvlei near Cape Town, clearly include part of the Late Pleistocene, but the radiocarbon dates suggest that at least part of the sequence is beyond the limit of the method. Schalke (1973) tentatively correlated the changes he observed in the pollen spectra with the Middle Pleniglacial (50 000–24 000 BP) sequence in north-west Europe, but no

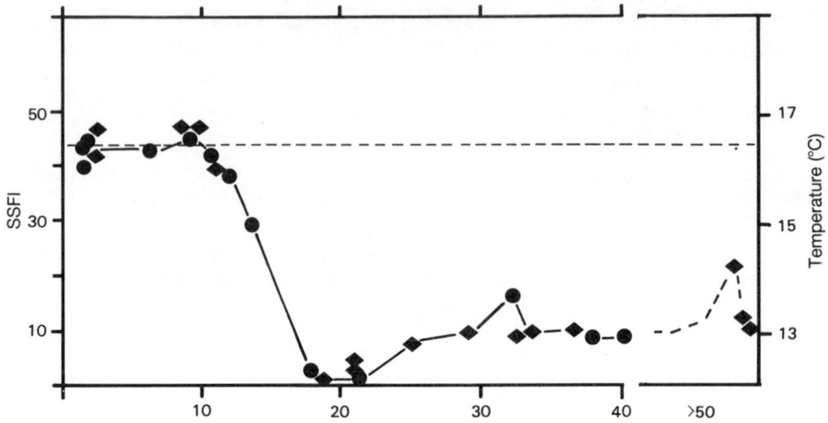

Fig. 6.7 Boomplaas Cave micromammal temperature curve. Relative changes in the value of a summary statistic calculated from factor scores of the relative frequency of micro-mammalian taxa (SSF1). Circles indicate samples dated by radiocarbon, diamonds are interpolated samples. The horizontal dashed line corresponds to modern SSF1 values and mean annual temperature (after Thackeray, 1987).

confirmation of this correspondence has been found in more recent assessments of the data (Scott 1984*a*). Van Zinderen Bakker (1976) has noted that, in subdividing the spectra into AP and NAP, Schalke has included the Proteaceae, Myrtaceae and most of the Rutaceae (which are generally regarded as bushy elements of the fynbos vegetation rather than trees) within the AP. The so-called 'forest elements', then, are not typical of the modern forest environments in the Cape Ecozone, and Schalke's interpretation of the sequence as demonstrating an increase in forest taxa is therefore spurious. There is indeed an increase in *Podocarpus* pollen, however, which may indicate that yellowwoods became more common in the vicinity between 45 000 and 40 000 BP, but there is no reason to assume a large-scale change to forest taxa on the Cape Flats at that time. Scott (1984*a*) notes that the range of taxa in these spectra suggests warmer and wetter conditions than those of the present which questions the precision of the dating and possibly gives grounds for placing this pollen zone further back in time than the radiocarbon dates indicate, i.e. to the Last Interglacial. The rest of the spectra from both Rietvlei and the Cape Flats sites show fynbos vegetation not very different from that of the present.

Near the coast at Groenvlei in the southern Cape in an area that was surrounded by Afromontane forest in historic times, five pollen zones have been recognized in a 7 m lake profile (Martin 1962, 1968) dating within the last 8000 years or more. The basal zone indicates somewhat restricted forest with fynbos vegetation reflecting relatively dry conditions with the lowest AP count in the sequence. This is below a date of *c.* 7000 BP (Martin 1968) and may in fact be considerably older—possibly as much as 12 000–15 000 years. Zone B2 shows a slight increase in AP suggesting slightly wetter conditions and a spread of forest around 6870 BP. Zone B3 represents a marine incursion of the lake probably coincident with the mid-Holocene rise in sea level to +1.5 m (Martin 1968) with slightly lower percentages of AP and more Restionaceae and Chenopodiaceae which Martin suggests were the result either of hotter, drier conditions or conditions similar to those of B2, but with greater mobility of sand in the local environment. Warmer conditions are confirmed by the range of marine diatoms deposited at this time (Martin 1968). Although *Podocarpus* pollen is at its lowest level in the next zone, C1, *Euclea* and *Sideroxylon* are prominent indicating a reduction of forest in favour of scrub elements around 2000 BP. The return of forest is indicated in C2 where *Podocarpus* counts are relatively high indicating more effective moisture.

In the same area as Groenvlei, Scholtz (1986) has sampled two organic peat deposits in the George district on the Norga and Gwayang rivers. The Norga sequence has a basal date of *c.* 3000 BP in one core, but a second undated core is assumed to date from *c.* 4000 BP. A cycle of forest advance and retreat is recorded in the sequence, beginning with expansion of forest between 4000 and 2600 BP and a wet local environment. Between about 2600 and 1400 BP, is a marked forest retreat with a decline in all forest taxa and the disappearance of *Myriophyllum*, a true aquatic. The forest and aquatic taxa are replaced by high percentages of grass pollen suggesting that grassland was prominent on these coastal forelands even prior to the advent of European farmers. After 1400 BP *Podocarpus* and *Myriophyllum* reappear and the forest expanded again. Recent ecological studies in the southern Cape by Van Daalen (1980) have concluded that the forests in this region must be seen as a relict of a much wetter climatic period and we can therefore suggest that the times of forest advance documented in the pollen record at both Groenvlei and Norga in the mid- and later Holocene indicate periods of higher effective precipitation. Because the Afromontane forest is particularly susceptible to summer drought, it may be that these periods of forest advance were triggered by an increase in summer rainfall rather than (or in addition to) an overall annual increase.

To the east of Groenvlei and Norga in the south-eastern Cape, ecological studies by Cowling (1983*a,b*) have suggested that here the most common vegetation type in the last 5000 years has been grassland rather than the South Coast Renosterveld suggested by Acocks (1953) and it would be of interest to test this hypothesis with palynological research.

In a somewhat different setting, where extensive forests were not known in historic times, a peat deposit from Cape Hangklip in the south-western Cape, with a basal date of 11 140 ± 65 BP (GrN-4586), fynbos vegetation is represented throughout the time period from about 6000 to 2000 BP with no changes of any significance. Pollen is not preserved throughout the time represented by the core, however, and changes may have occurred without being recorded (Vogel 1970; Schalke 1973; Scott 1984*a*).

One other pollen sequence from the Cape Ecozone, that from Noordhoek on the Cape Peninsula, has an undated Quaternary spectrum that overlies a much longer sequence recording vegetation changes from Late Oligocene through Miocene and Pliocene times. The Quaternary pollen zone, M, is characterized by fynbos vegetation (Coetzee 1978).

Non-archaeological charcoals recovered from a soil pit in indigenous forest at Kaffirkop, east of Knysna on the southern Cape coast in the same general area as Groenvlei, have been identified as *Olea* cf. *capensis macrocarpa*, a forest species (Deacon *et al.* 1983). The sample has been radiocarbon dated to 8000 BP (J. C. Vogel, personal communication), indicating that Afromontane forest was established on the outer margin of the coastal platform at this time and suggesting therefore that the dry heath vegetation preceding the forest expansion noted in the early Holocene pollen at Groenvlei could be somewhat older than 8000 BP as originally suggested by Martin (1968).

In the Cedarberg mountains of the western Cape 32 trees (*Widdringtonia cedarbergensis*) were examined in a pilot study of the potential for dendrochronology in this region (Dunwiddie and LaMarche 1980). The recent tree-ring record, compared against modern climatic data for the same period, shows a complex response to climatic variables, but 73 per cent of the year-to-year variability in tree ring width can be explained by departures from normal temperature and precipitation. Dunwiddie and LaMarche (1980) suggest that the tree ring chronology represents primarily a record of spring and early summer moisture availability, with other factors such as severe frosts in 1815 (linked to the unusually large volcanic eruption at Tambora at 8° south), and wide-spread forest fires also playing a part. As such, the data would seem not to be a reliable and easily interpretable record of annual rainfall or temperature variations.

6.8 Faunal changes in the Cape Ecozone

Klein (1980, p. 252) has pointed out that although the Cape Ecozone is the smallest of the Southern African ecozones, it has the largest sample (31) of well excavated and described Late Pleistocene and Holocene larger mammal faunal assemblages. They date from the Last Interglacial, Last Glacial, and Present Interglacial, and come from both archaeological and non-archaeological contexts in all regions of the ecozone.

Large mammal faunal samples dating to the Last Interglacial are more common than in other ecozones, but are nevertheless rare and scattered observations. The majority (six out of nine) were accumulated by carnivores (mostly hyaenas) rather than people, and those that are archaeological are associated with Middle Stone Age artefacts (Klein 1980). Both types of site have faunal assemblages dominated by browsing ungulates, but the inclusion of the range of 'giant' forms that became extinct at the end of the Pleistocene makes the community structure of these large mammals rather different from that of the Present Interglacial. With the onset of cooler conditions at the start of the Last Glacial, grazers become more common (Klein 1980) until they dominate all assemblages through the Last Glacial Maximum, as noted above, and were replaced by browsers again during the Holocene.

Although samples from Holocene sites do not show change at the same scale as that seen between the Late Pleistocene and Holocene, there are nevertheless some shifts in the relative frequency of certain ungulates which are of interest. The early Holocene assemblages dating between about 10 000 and 5000 BP have relatively high numbers of vaalribbok, mountain reedbuck and roan antelope, whereas these species are much rarer in the assemblages post-dating 5000 BP in which the blue duiker first appears. The implications of this shift are not clear as yet (Klein 1980, p. 259), but it may be linked to vegetation changes noted in the pollen data summarized above.

Size variation in mammals from the Cape Ecozone has been a useful indicator of past temperature and rainfall. Size estimates have been made by Klein (1984) on jackal, dune molerat, hyrax, and/or grysbok bones from Elands Bay Cave, Tortoise Cave, Sea Harvest, Swartklip, Die Kelders, Byneskranskop, Nelson Bay Cave, and Klasies River mouth. Overall, they suggest that the Last Interglacial had alternating periods of warm/wet, warm/dry and cold/wet conditions. The Last Glacial was initially cold/wet, whereas the Last Glacial Maximum was cold and very dry. The early/mid-Holocene was warmer and drier than the later Holocene.

This generalized picture is not entirely consistent with the changes in size of all the mammals at all sites, however. Modern dune molerats, *Bathyergus suillus*, are smaller in areas of low rainfall (for example near the mouth of the Orange River) than they are in areas of higher rainfall in the southern Cape. Klein (1984, p. 141 ff) has shown that there are variations too in Late Pleistocene and Holocene samples from Tortoise Cave, Elands Bay Cave, and Sea Harvest in the drier western Cape and from Die Kelders 1 and Byneskranskop 1 in the south-western Cape. The largest molerats come from the Middle Stone Age units at Die Kelders where they are on average 10 per cent larger than their counterparts from the Late Holocene units. This confirms conclusions drawn from the sediments (see above) that also indicated higher effective precipitation at that time.

Amongst carnivores, Klein (1983, 1986) has used carnassial length as an indicator of body size and notes that length tends to increase with latitude south in the black-backed jackal, Cape fox, hunting dog, Cape otter, water mongoose, Egyptian mongoose, grey mongoose, spotted hyaena, brown hyaena, wildcat, serval, caracal, lion, and cheetah. Carnassial length tends to decline with latitude south in the striped polecat and honey badger, and is probably not related to latitude in the leopard. Klein (1986) has shown further that, with the exception of the honey badger, fossil carnivores that can be dated to known cooler intervals are significantly larger than their modern counterparts.

The black-backed jackal, *Canis mesomelas*, is the most sensitive to

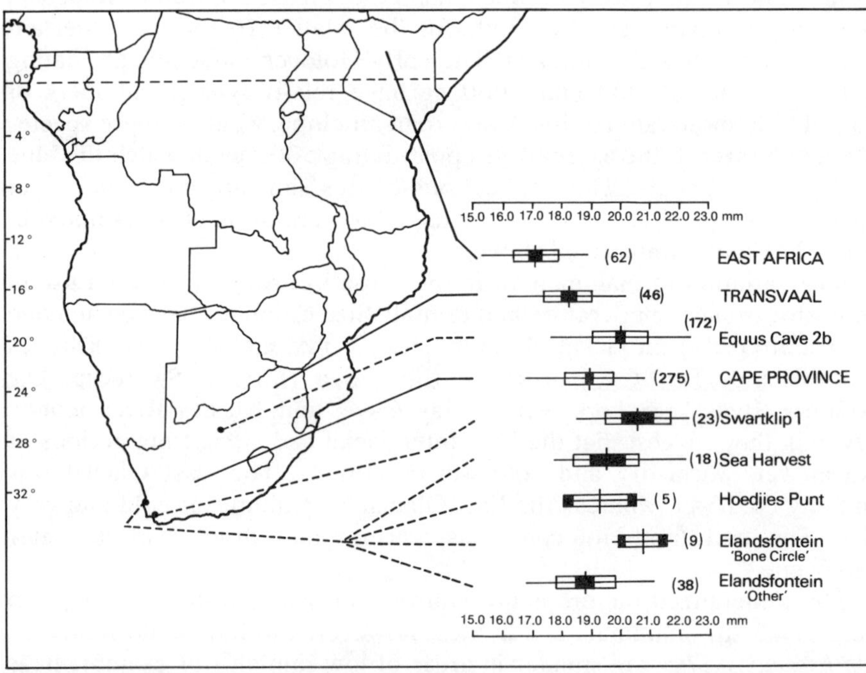

Fig. 6.8 Location and measurements of recent and fossil first molar (M_1) of the black-backed jackal, *Canis mesomelas*. The modern samples (upper case) show that individuals closer to the equator tend to be smaller than those further south. The implication is that larger individuals in fossil samples (lower case) indicate cooler conditions (after Klein 1983).

temperature of the carnivores measured by Klein (Fig. 6.8). The lower carnassial teeth, taken to reflect body size, increase approximately 0.08 mm in average length per degree latitude south. The fossil sample from Swartklip near Cape Town, with a mean lower carnassial length of 20.58 mm (compared with modern samples of about 16.70 mm at the present-day equator), suggest a mean annual temperature of 12°C, or about 5°C cooler than the mean annual temperature at Swartklip today. With a radiocarbon date on ostrich eggshell of greater than 40 000 BP, Klein (1986) has suggested that the sample dates to a cool interval pre-dating the Last Glacial Maximum, possibly within oxygen isotope stage 5. This would make the temperature depression similar to that measured for the Last Glacial Maximum at Cango and the Uitenhage aquifer in the southern Cape.

6.9 Isotopic evidence from the Uitenhage aquifer and Cango speleothem

Direct quantitative calculations have been made for palaeotemperatures in the past 45 000 years at two sites in the southern Cape giving some estimates for the scale and timing of temperature changes in the sub-continent.

Fossil water from the Uitenhage artesian aquifer north-east of Port Elizabeth has been sampled from the area of recharge for about 30 km along the aquifer (Heaton 1981; Vogel 1983; Heaton *et al.* 1986*a*). The water represents rainwater that filtered through the quartzite outcrop at the northern end of the aquifer and slowly flowed underground through the artesian basin towards the sea. Radiocarbon dates show that the water nearest the coast is about 28 000 years old.

Heaton (1981; Heaton *et al.* 1986*a*) has measured the concentration of nitrogen-2 and argon in 21 dated samples of the groundwater. Gas solubilities are temperature-dependent, and once rain water has been isolated from the atmosphere in an underground aquifer, no further change takes place. The concentration of atmospheric gases dissolved in groundwater is therefore determined by the temperature of the water-table at the time of groundwater recharge. The samples dating between 28 000 and 15 000 BP showed mean temperatures at recharge ranging between 12 and 16°C, with a mean of 14 ± 1°C. As present temperatures are in the range of 18–21°C, this means a 5.5°C drop in mean temperature at Uitenhage during the Last Glacial Maximum. The samples dating between 9000 BP and the present ranged from 16 to 23°C, with a mean of 19.5 ± 2°C (Heaton *et al.* 1986*a*; Fig. 6.9).

The oxygen-18/oxygen-16 ratio of the groundwater has also been

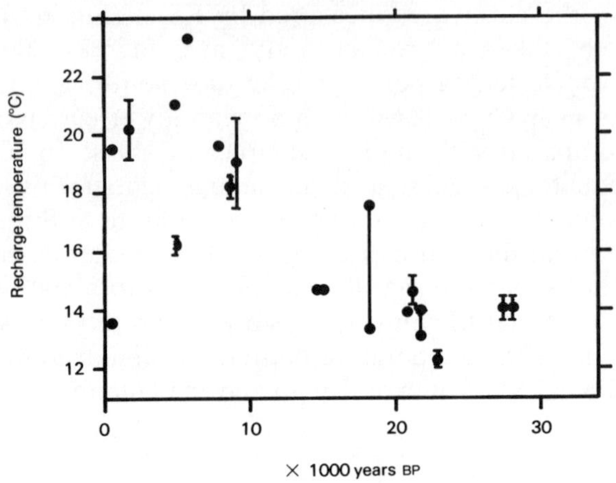

Fig. 6.9 Palaeo-recharge temperatures for the Uitenhage aquifer groundwater. Means and ranges are shown where temperatures were derived from duplicate analyses. Individual values connected by tie lines are shown for the two boreholes sampled on separate occasions (after Heaton *et al.* 1986*a*).

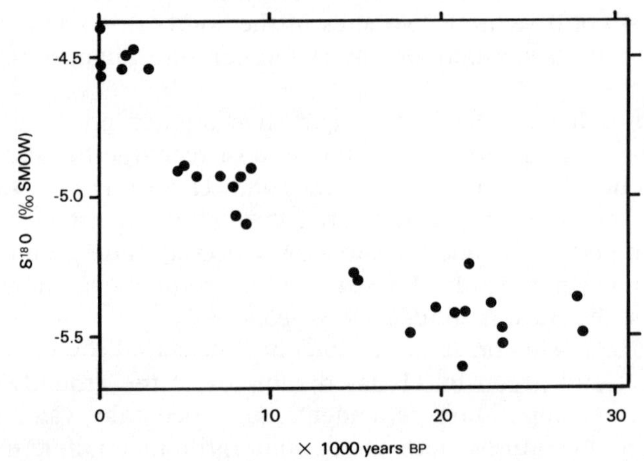

Fig. 6.10 Oxygen-18 variation with groundwater age in the Uitenhage aquifer (after Heaton *et al.* 1986*a*).

measured (Heaton *et al.* 1986*a*). The youngest samples representing the most recent recharge have a characteristic oxygen-18 of −4.5‰, whereas those older than 15 000 BP have an average of −5.4‰ and those of intermediate age (4800–900 BP) have an average of −5.0‰ (Fig. 6.10). These changes in the isotopic composition of precipitation are related partly to changes in temperature, but there are uncertainties involved in the estimation of accurate palaeotemperatures. Vogel (1983) calculated that temperatures between 15 000 and 28 000 BP had been 5.5°C cooler than the present, but Heaton *et al.* (1986*a*) point out that a temperature difference of 8°C could also be derived from the same data using slightly different, but still acceptable, assumptions.

Vogel (1983) reports oxygen isotope changes in a radiocarbon dated section through a 2.8 m speleothem from a sealed underground cavern in the Cango Caves in the Swartberg north of Oudtshoorn. The almost continuous sequence dates from *c.* 45 000 to about 15 000 BP, there is a break in deposition to the mid-Holocene, and then the stalagmite grew again continuously to modern times. The oxygen isotope ratio in the calcite depends on that of the groundwater dripping into the cave and the temperature at which the carbonate is precipitated from this water. Vogel (1983) has reasoned that if it is accepted that a shift of 6‰ in the deuterium content corresponding to 0.8‰ in oxygen-18, as observed in the Uitenhage groundwater, applies generally to the southern coast of the Cape Province, then temperatures of crystallization for the Late Pleistocene can be calculated. The results suggest that the coldest temperatures occurred about 17 000 BP and that the Last Glacial Maximum temperature was 5°C cooler than at present (Fig. 6.11). The Holocene record suggests a range of about 1°C around the present day mean after *c.* 5500 BP (Fig. 6.12).

A. S. Talma and J. C. Vogel (personal communication) have also measured the carbon-13 content of the speleothem over the past 47 000 years that reflects, amongst other factors, the nature of the vegetation through which the rain water passed before entering the cave. The results show (Fig. 6.13) that there is a clear difference between samples dating to the Last Glacial that indicate C_3 plants in the vicinity, and Holocene samples that show a relatively late shift to C_4 plants. The meaning of the

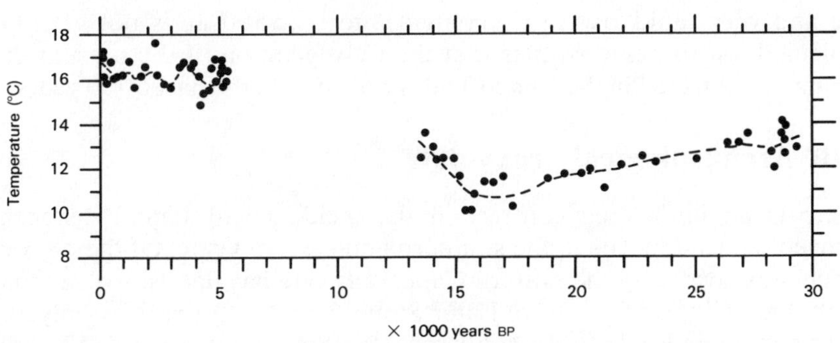

Fig. 6.11 Isotopic temperatures for the past 30 000 years derived from Cango Caves speleothem V3. Results show a clear contrast between Last Glacial and Present Interglacial values and a mean difference of 5.5°C between them (after Vogel and Talma, personal communication).

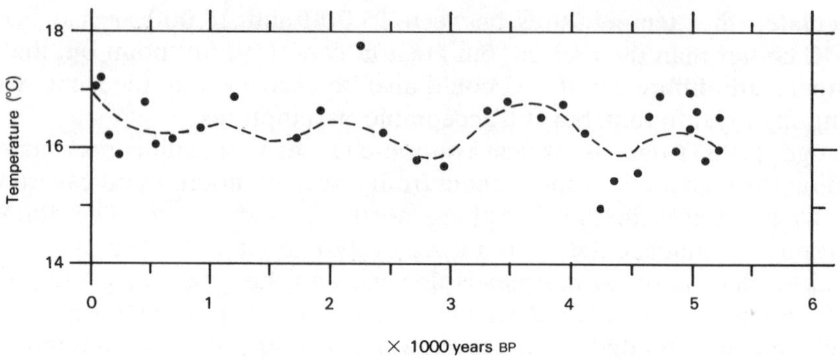

Fig. 6.12 Isotopic temperatures for the last 5500 years derived from the younger portion of Cango Caves stalagmite V3. Results show variation of 1 degree C around the present day mean (after Vogel and Talma, personal communication).

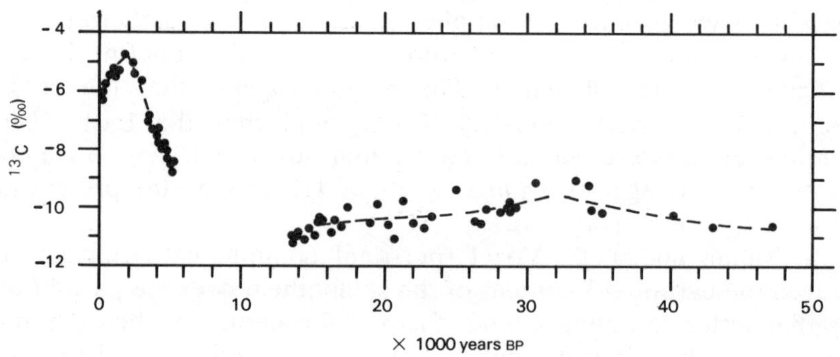

Fig. 6.13 Spline curve of carbon-13 variation of Cango Caves speleothem V3 during the last 47 000 years. Carbon-13 values in the −6 to −5 range are typical of present-day C4 vegetation, whereas values of −11 to −9 are typical of modern C3 vegetation in cooler regions of South Africa (after Vogel and Talma, personal communication).

late and very rapid rise in C_4 content after *c.* 5500 BP is not yet fully explained, but it demonstrates that the C_4 vegetation of the present day became established in the Cango Valley only within the past 5000 years.

6.10 Archaeological proxy data

There is an interesting contrast in the incidence of Late Pleistocene archaeological deposits in the southern and eastern Cape. Of the 18 rock shelters excavated in the eastern Cape, only one has deposits older than 15 000 BP; all the others have Later Stone Age (LSA) deposits only and these rest on bedrock. In the southern Cape, on the other hand, 13 of the 19 excavated sites have MSA material underlying the LSA. The western Cape sites are intermediate between the southern and eastern Cape ones with fewer long-sequence sites occurring north of Cape Town than occur between the Cape Peninsula and the Gouritz River. Deacon and Thackeray (1984) have suggested that this is proxy evidence for greater environmental productivity in the southern and south-western Cape during the Late Pleistocene, implying that faunal and vegetation changes during the Last Glacial were more favourable to hunter-gatherers there than was the case to the north and east. The temporal distribution of sites also stresses the dramatic increase in diversity and productivity after 15 000 BP throughout the Cape Ecozone, a factor that was possibly linked also to increased efficiency of hunting equipment and gathering strategies.

6.11 Chronological summary

The Cape Ecozone has a more complete record of climatic change interpreted from biological data than any other ecozone in southern Africa. This is partly because of active research programmes over the last 50 years, partly because environmental conditions in the region were apparently attractive to hunter-gatherers for long periods in the Late Pleistocene and Holocene, and partly because conditions for the preservation of bone have been favourable at a number of long sequence sites. Interdisciplinary research at Boomplaas and other sites has meant that rates of change in larger mammal and micromammal faunas can be compared against those observed in pollen spectra and charcoals, and more isolated observations at other sites can be compared against this sequence. The location of several sites on the coast (Klasies River Mouth, Herolds Bay, and Die Kelders) has helped in the dating of the deposits relative to changes in sea level.

The best example of Last Interglacial deposits comes from the Klasies River Mouth caves on the southern Cape coast where the lowest units are estimated, on the basis of the associated larger mammal fauna, to date to the Last Interglacial with a combination of species, dominated by browsers, similar to that found in the area in historic times. In the overlying deposits, dating between about 125 000 and about 70 000 BP, there is evidence for a general cooling of temperatures which resulted in the replacement of bushy vegetation with a vegetation mosaic that included substantially more grass—enough to sustain herds of migratory grazers such as equids and alcelaphines, including the giant buffalo, a giant alcelaphine, and the large Cape horse, all of which became extinct at the end of the Pleistocene (Klein 1976, 1980). The range of micromammalian taxa confirms this reconstruction of the vegetation (Avery 1982*a*). Cool and wet conditions in the Howiesons Poort levels (dated by the aspartic acid method to 89 000 BP) are indicated by the highest incidence of equids and alcelaphines in the sequence, by the micromammalian taxa represented, and by the oxygen isotope content of a marine shell (Shackleton 1982; Singer and Wymer 1982). On the basis of these changes in temperature inferred from the fauna, the sequence has been fitted into the period from stage 5e to the beginning of stage 4 in the deep sea oxygen isotope sequence (Butzer 1978; Singer and Wymer 1982; Deacon *et al.* 1986).

Inland, the base of the Boomplaas sequence includes artefacts typical of the Howiesons Poort (i.e. at least 70 000 BP) that are overlain by regularly spaced samples through to the end of the Holocene. The large and small mammals, charcoals and pollen all confirm that the period between 80 000 and about 25 000 BP was distinctly cooler than the present, but not as cold as in the Last Glacial Maximum. There is also some evidence from the charcoal and micromammalian data for moister periods, but the larger mammal fauna shows little variation, being dominated throughout by equids and alcelaphines with vaalribbok common at times (Klein 1978*b*; Deacon *et al.* 1984).

The coldest and driest conditions of the last 125 000 years date to the Last Glacial Maximum from 20 000 to about 16 000 BP. The vegetation at that time was again markedly grassier and more open than at present with a low species diversity in charcoals, pollen and micromammals. Confirmation of drier conditions at the Last Glacial Maximum is given in the micromammals from Nelson Bay Cave, at present on the southern Cape coast (Avery 1982*a*), and in the mean individual size of small herbivores (Klein 1984).

In the period between the end of the Last Glacial Maximum and the beginning of the Holocene 10 000 years ago, dramatic changes can be seen in the biological data that relate to the relatively sudden rise in global temperatures and the subsequent adjustment of plant and animal communities. This shift is more dramatic in the Cape Ecozone than elsewhere and was initiated before 14 000 BP, confirming observations in other southern hemisphere continents that the post-glacial warming took place some 2000 years or more earlier than in the northern hemisphere (Salinger 1981, 1984; Labeyrie *et al.* 1986). Apart from being somewhat warmer, the evidence from Boomplaas suggests that this period was also much wetter than in either earlier or later times. The response of the vegetation, noted both in the charcoals and the micromammals, was to increase the number of trees and shrub taxa at the expense of grass, yet it was not until after 12 000 BP that there was a marked change in the range of larger mammals and a shift towards small non-gregarious browsers. This apparent anomaly stresses the fact that various biological proxy indicators of climatic change (pollen, charcoals, micromammals, larger mammals, shellfish) differ in the rate and nature of response to climatic change, the vegetation and micromammals being more immediately affected than the larger mammals (Deacon *et al.* 1984*a*). Technological change in the stone and other artefacts made by Stone Age people also appears to lag behind vegetation and micromammalian shifts (Deacon and Deacon 1986). At Boomplaas this time lag was of the order of 2000–3000 years.

The rate of change observed in sea level rises off the Southern African coast and worldwide is a good indicator of the speed of atmospheric warming in the southern hemisphere during the terminal Pleistocene (Fig. 6.3). Southern African dates are drawn from littoral shells and an estimate of the degree of ocean cooling at the Last Glacial Maximum to 5°C lower than at present is derived from the study of planktonic foraminifera in the Mozambique Channel (Vincent 1972). The Uitenhage aquifer and Cango speleothem confirm this scale of change on land as well. By 10 000 BP sea surface temperature was similar to that of the present. Early Holocene changes in sea temperature can also be seen in the shift in species frequencies of shellfish at Nelson Bay Cave which alter to modern community composition at *c.* 11 000 BP (Klein 1972*a*).

All data examined from the southern Cape show that the last 10 000 years have been markedly warmer than at any other time during the last 100 000 years, although a marginally cooler interval is suggested from both micromammal and pollen studies in the last 2000 years (Martin 1968; Avery 1982*a*; Scholtz 1986; Thackeray, in press) (Fig. 6.7). All Holocene larger mammal faunas are dominated by browsers in this ecozone although there was a change in emphasis after 5000 BP (Klein 1980), and the micromammalian community structure was distinctly different at all sites from that observed during the Last Glacial cycle (Avery 1982*a*). Pollen studies from the southern Cape seem to indicate two periods of somewhat warmer conditions between 7000 and 2600 BP when forest elements increased in importance in most pollen spectra (Martin 1968; Scholtz 1986). Thereafter, a decline in forest taxa and an increase in grassland species shows that in the southern and south-eastern Cape at least, grassland was the 'natural' climax prior to the present-day renosterveld with forest increasing only in warmer times when rainfall, particularly in summer, was also higher. The linking of these periods of forest expansion to the mid-Holocene rise in sea level to +1.5 m above that of the present day has not been possible because of the changes in local soils and topography with the movement of dune sands at Groenvlei, the only coastal site in the forested area where pollen of this age is preserved.

7 Summary and conclusions

Reliable information on the climatic history of Southern Africa over the last 130 000 years has not been easily won and data collection has inevitably been influenced by prevailing hypotheses. Models based on the European glacial sequence and the pluvial hypothesis sought data on past rainfall changes from the sedimentological and geomorphological record. By the time this model was dropped in favour of one based on the equatorward movement of climatic belts during glacial cycles, biological data were available, and both temperature and rainfall changes could be traced. In the past few decades, deep sea cores have given more detail on global sea temperatures, but in Southern Africa this record is in need of systematic testing in land-based sequences. Isotope records in speleothems and aquifers are being used for this purpose and information is being sought on the response of geomorphic, sedimentary, and biological systems to the periodic changes in climate during the Quaternary. The results summarized here give an idea of how much, and how little, we know.

The radiocarbon dates that relate to palaeoclimatic data described in this book are summarized in Appendix 1. Figures 7.1, 7.2, and 7.3 use radiocarbon dated occurrences that have given reliable *primary* palaeoclimatic data as defined by Hecht *et al.* (1979) (see chapter 1.1). In the summary that follows, the evidence for climatic change during the past 130 000 years is discussed on an inter-regional basis. The time periods are loosely linked to the oxygen isotope chronology, but do not imply a formal correlation.

7.1 *c.* 130 000–80 000 BP

The most securely dated site in this time range is the Cave 1/1a complex at Klasies River Mouth on the southern Cape coast (Singer and Wymer 1982). The Middle Stone Age (MSA) deposits have accumulated on the floors of Caves 1 and 1b at 6–8 m above present sea level. The caves were formed by several Pleistocene high sea levels and the pebbly beach at the back of Cave 1 probably relates to one of these (Deacon *et al.*

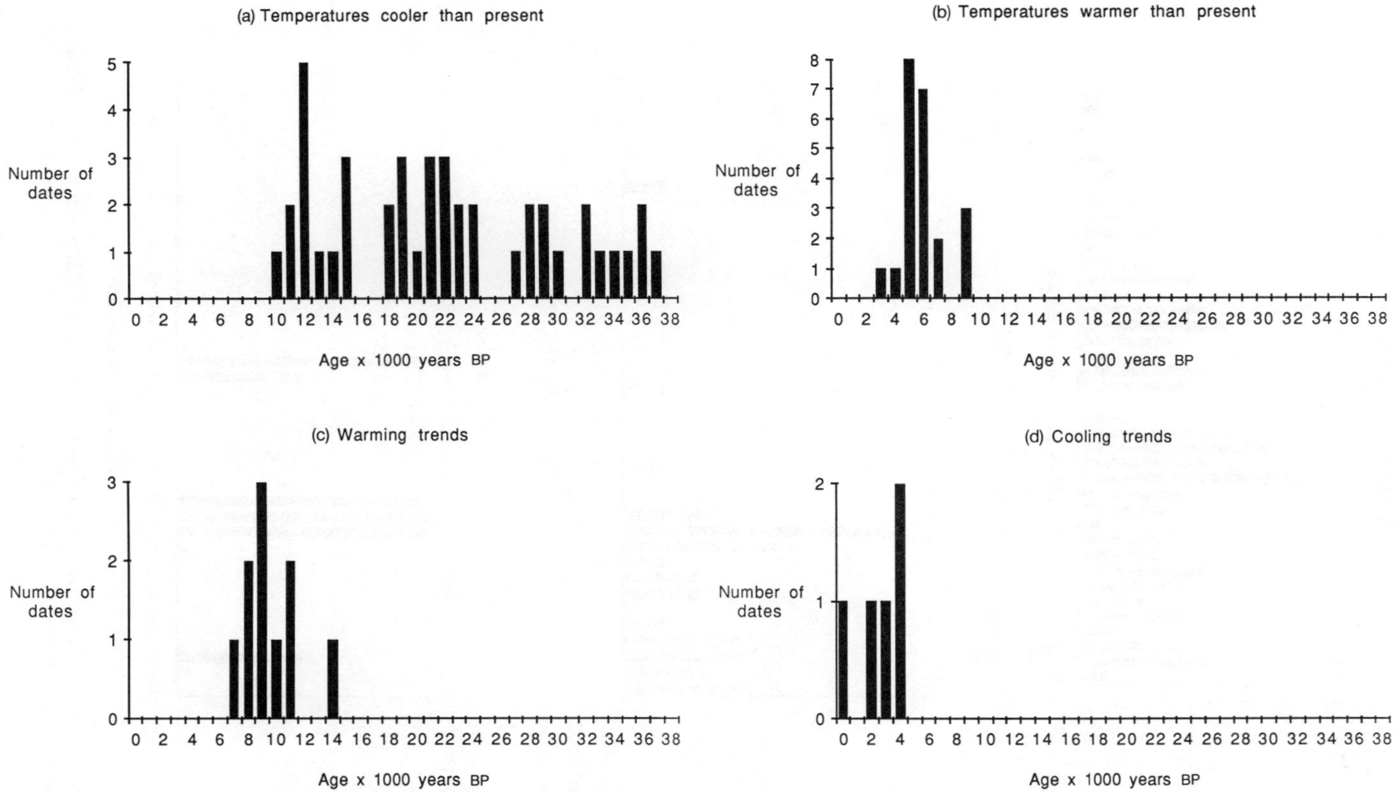

Fig. 7.1 Radiocarbon dated instances of past temperature conditions in Southern Africa (a) cooler than at present, (b) warmer than at present, (c) showing a warming trend, i.e. conditions warmer than the previous period, but not as warm as at present, and (d) showing a cooling trend, i.e. cooler than the previous period.

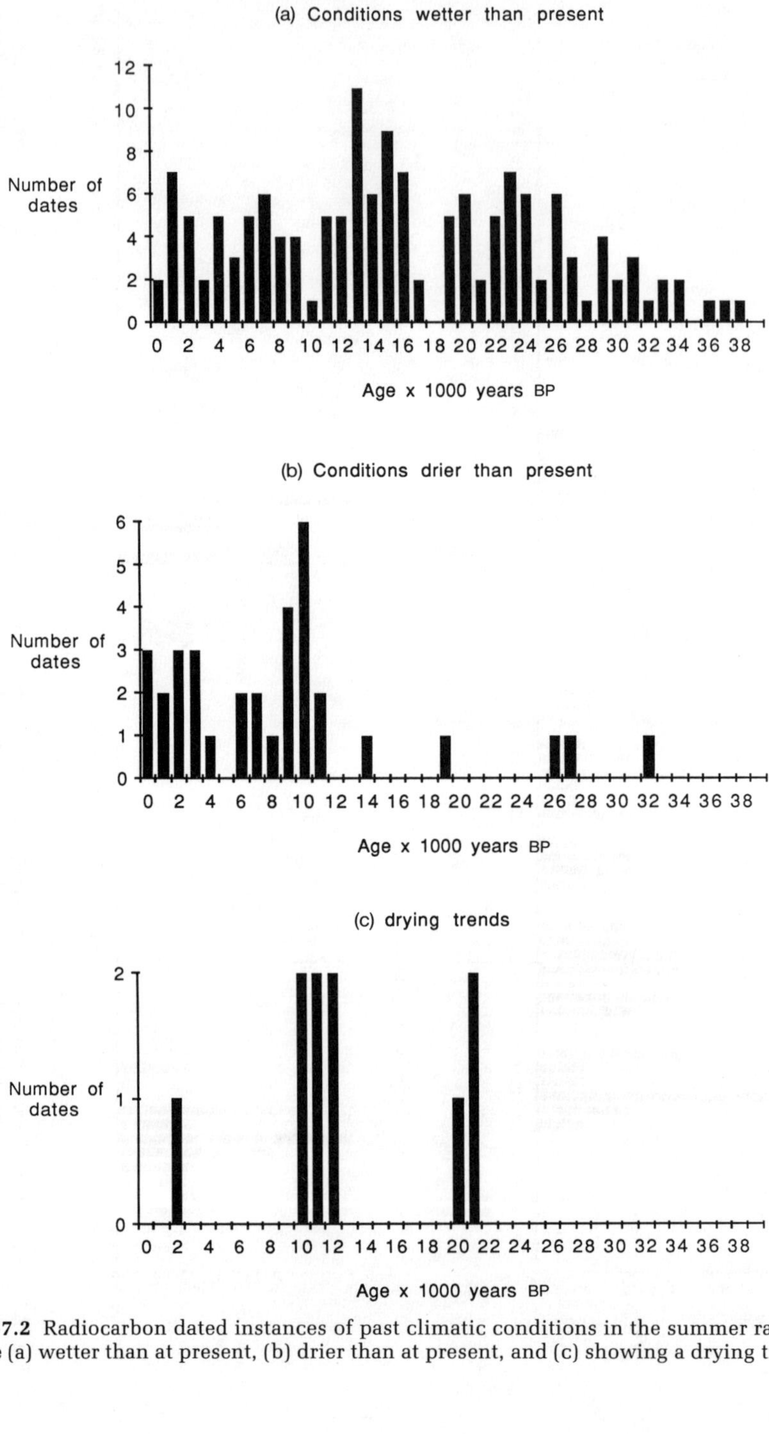

Fig. 7.2 Radiocarbon dated instances of past climatic conditions in the summer rainfall zone (a) wetter than at present, (b) drier than at present, and (c) showing a drying trend.

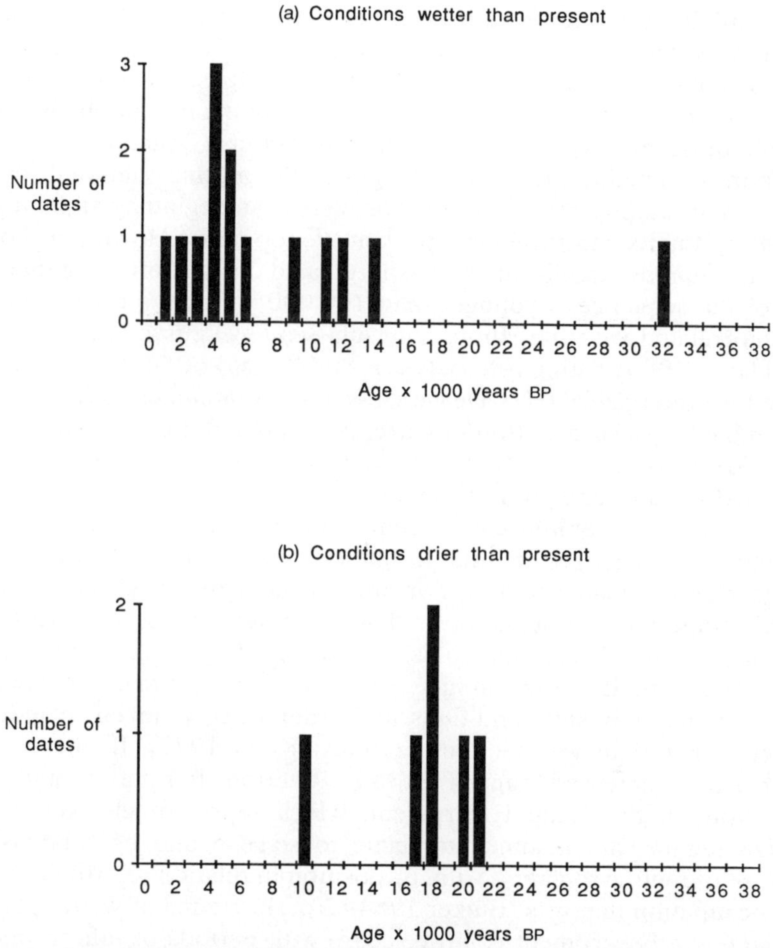

Fig. 7.3 Radiocarbon dated instances of past climatic conditions in the winter and all-year rainfall zone (a) wetter than at present, and (b) drier than at present.

1986). The large mammals and shellfish species in the lowermost (MSA I and MSA II) deposits are similar to those of the Present Interglacial suggesting that they either coincide with or immediately post-date the warmest phase of the Last Interglacial. In the overlying units (Howiesons Poort, and MSA III and IV) the fauna is dominated by grazers rather than browsers which have been shown to characterize most of the Last Glacial cycle in the Cape Ecozone (Klein 1980, 1983) and indicate cooler temperatures which promoted grasslands rather than the bushier environments typical of interglacials. The oxygen isotope ratios of marine shells (Shackleton 1982; H. J. Deacon, personal communication) show initially warm temperatures in the MSA I and II with an isotopic signature typical of the Last Interglacial (stage 5e) at the base and cooler, but not cold temperatures in the overlying samples. Sub-stages within stage 5 are recognizable in the cave sediments, in the marine shells and in the micromammalian taxa. The recent excavations (Deacon *et al.* 1986) give a more detailed picture of changes in the marine shell and micromammalian samples that show a major regression beginning at the top of the MSA II units and reaching a peak at the top of the Howiesons Poort. With uranium disequilibrium and aspartic acid dates confirming that this part of the sequence is younger than 100 000 years, the regression has been correlated with the Huon Peninsula sea level curve (Chappell and Shackleton 1986) dating it to between 80 000 and 60 000 BP at the end of the Last Interglacial (H. J. Deacon, personal communication).

At inland sites such as Border Cave, Alexandersfontein, Gaap Escarpment, Rose Cottage, and the Lower Vaal River terraces, there are either no faunal remains and/or no formations such as raised beaches to tie the observations directly into the oxygen isotope sequence. Instead, it is the alternation of changes in tufa formation, sediments, and the species frequencies of small and larger mammals, interpreted as the result of gross temperature fluctuations, that is assumed to represent the responses to the warmer and cooler sub-stages of oxygen isotope stage 5. Unlike the Cape Ecozone, an increase in browsers and micromammalian taxa that prefer bushy conditions at Border Cave is taken to indicate cooler rather than warmer temperatures (Klein 1977). In the case of other sites, increased run-off in the Riverton Formation and tufa formation on the Gaap Escarpment, which represent changes in the rainfall regime, are assumed to relate to stage 5 and its subdivisions because of their association with, or position in relation to, Middle Stone Age occupation deposits (Butzer 1984*a*,*b*). The dating of wetter periods on the Gaap Escarpment is more secure with periods of tufa formation dated by thorium to 103 000 ± 6000 BP and by uranium to 86 000 ± 21 000 BP (Vogel and Partridge 1984, p. 512), but the standard errors are large. There are no well dated deposits with palaeoenvironmental data from the winter rainfall region to compare with those to the north and east.

7.2 *c.* 80 000–50 000 BP

This time period is still beyond the resolution of radiocarbon dating and the assignment of deposits to it depends very largely on their stratigraphic position in long sequences or on the associated stone artefacts.

The most complete sequence comes from Boomplaas Cave in the southern Cape with additional data from Die Kelders and Klasies River in the Cape Ecozone and Border Cave in the Transvaalian Ecozone. At all three, biological data and cave sediments suggest cooler conditions than the preceding time period. The faunal remains from the Cape sites all indicate grassland rather than bushy vegetation, and at Boomplaas the relatively high incidence of *Pelea capreolus*, the vaalribbok, indicates cool temperatures as does the range of micromammalian species (Klein 1976, 1978*b*; Avery 1982*a,b*; Deacon *et al.* 1984*a*). Temperatures derived from factor scores on Boomplaas microfauna suggest at least one warmer peak within cool conditions in this general period (Thackeray, in press). At Border Cave the fauna is represented by taxa which prefer bushier environments (Klein 1977; Beaumont *et al.* 1978; Avery 1982*b*), and at Redcliff in Zimbabwe cooler temperatures are indicated by the extension of the range of some larger mammals which only occur further south today (Cruz-Uribe 1983). Cold intervals when temperatures were low enough to induce frost spall horizons in cave sediments may be dated to this time range at Melikane, Rose Cottage, Bushman Rock Shelter, Border Cave, and several sites in the southern Cape, for example Nelson Bay Cave, Boomplaas, Diepkloof, Die Kelders and Klasies River (Carter 1976; Butzer 1984*a,b*; Deacon *et al.* 1984).

Moister conditions are apparently indicated by tufa formation on the Gaap Escarpment dated by thorium to 61 000 ± 5000 BP (Vogel and Partridge 1984, p. 512). Undated charcoal samples from this time range at Boomplaas and micromammals at Die Kelders both suggest it was drier (Avery 1982*a*; Scholtz 1986) and the Boomplaas wood anatomy is suggestive of windy conditions as well (Scholtz 1986).

7.3 *c.* 50 000–24 000 BP

The oxygen isotope record in both the Wolkberg and Cango caves stalagmites begins within this time period. Wolkberg, with fewer readings, shows a gradual cooling from about 40 000 to about 20 000 BP (Talma *et al.* 1974). Cango shows a slight cooling trend from *c.* 45 000 to 30 000, and then a cool but fairly stable record up to about 22 000 BP (Vogel 1983). The carbon-13 content of the Cango stalagmite also confirms cooler temperatures over this time range with C_3 plants predominating (Vogel and Talma, personal communication). The fossil water from the Uitenhage aquifer dates from about 28 000 BP and also shows fairly stable temperatures from the beginning of the sequence up to about 20 000 BP (Heaton *et al.* 1986*a*). At Boomplaas in the charcoal and micromammal data there is the suggestion of a slightly warmer interval centered around 32 000 BP (Figs 6.5 and 6.7) (Deacon *et al.* 1984*a*) that is possibly present also in the carbon-13 record from Cango Caves (Vogel and Talma, personal communication). By contrast, frost spalling is reported from Border Cave between 38 000 and 33 000 BP, from Wonderwerk Cave between *c.* 30 000 and 26 000 BP, and at several cave sites in Lesotho, including Melikane (Carter 1976; Butzer *et al.* 1978*a*; Butzer 1984*b*). Mechanical weathering on the Gaap Escarpment before the formation of Tufa II is interpreted by Butzer (1984*b*) as indicative of increased frost and moister conditions around 30 000 BP.

In several ecozones this period is characterized by the wettest conditions of the past 130 000 years. Maximum lake levels were reached in the Kalaharian and Karoo-Namaqualian Ecozones. The Makgadikgadi lake levels are reported to have been markedly higher at various intervals between 40 000 and 20 000 BP (Cooke 1980, 1984; Heine 1978, 1982). Higher water tables in interdune areas in the Namib and stabilized dunes in the south-western Kalahari, both indicating much moister conditions, are dated to between about 32 000 and 28 000 BP (Heine 1982; Teller and Lancaster 1985). Cave sinters at Rössing between *c.* 41 000 and 26 000 BP also suggest increased humidity (Heine 1984; Heine and Geyh 1984). In the Bushmanland area of the Karoo-Namaqualian Ecozone, shell deposits have been interpreted as evidence for higher water levels at Breek Been Kolk between *c.* 38 000 and 14 000 BP (Kent and Gribnitz 1985; Beaumont 1986). At Wonderkrater in the Transvaalian Ecozone higher precipitation is indicated by the expansion of woodland and forest (Scott 1982*a*).

In the Cape Ecozone in the all-year rainfall region, more effective precipitation is suggested by the charcoals from Boomplaas Cave which include a larger number of woodland taxa (Deacon *et al.* 1984) and by the high incidence of *Myosorex varius* amongst the micromammals (Avery 1982*a*, 1983). Xylem analysis of the charcoals suggests it was cold with relatively high winter rainfall (Scholtz 1986).

Northern hemisphere palaeoenvironmental data summarized by Kukla *et al.* (1981) and oxygen isotope data from Indian Ocean deep sea cores (Chappell and Shackleton 1986) suggest that we may expect land-based sites dating to this time period to have been cooler than in the preceding 30 000 years, but there is little clear evidence for this in Southern Africa. Widespread evidence for moister conditions argues against severely cold temperatures, and while it was still cool enough to sustain grassland rather than browse in the southern Cape, and more C_3 plants grew in the highlands of Lesotho and the southern Cape than occur there at present (Vogel 1983, personal communication), the larger mammal fauna in parts of the interior (particularly around the eastern Transvaal at Heuningneskrans) was essentially similar to that of today. The same range of larger mammals occurs in the human occupation layers of that time as occurred during the Holocene levels at the same sites, but this may merely emphasize the fact that the larger mammals in the eastern part of the summer rainfall region are not a particularly sensitive indicator of past climatic changes.

7.4 *c.* 24 000–16 000 BP

Without exception, all palaeoclimatic data relating to this time period concur that it was the coldest interval of the last 130 000 years. Estimates of the temperature reduction vary from 8–9.5°C at Wolkberg in the Transvaal (Talma *et al.* 1974) to 5.5 and 5°C lower than at present in the Uitenhage aquifer and the Cango Caves speleothem respectively (Vogel 1983; Heaton *et al.* 1986*a*).

For the most part conditions were also drier, as predicted in general circulation models. In the Namib and Kalahari evidence points to falling lake and groundwater levels culminating in dry conditions around 19 000–18 000 BP (Heine 1978; Helgren and Brooks 1983; Teller and Lancaster 1985). The end of the long-continued sequence of high lake levels is marked by a relatively large number of radiocarbon dates for this event and should be interpreted as an indication of the onset of drier conditions rather than a marked wet period. Drier conditions are reported also from Haaskraal in the Karoo (Partridge and Dalbey 1986) and Kathu Pan in the southern Kalaharian Ecozone (Beaumont *et al.* 1984). To the east in the Basutolian Ecozone, pollen analyses suggest cooler and drier conditions at Cornelia and Craigrossie between *c.* 20 000 and 12 000 BP (Scott 1986). In the winter rainfall region of the western Cape, cool, dry, and windy conditions are evidenced by sediments in Elands Bay Cave (Butzer 1979) and by the size of molerats dating to some time between *c.* 20 000 and 11 000 BP (Klein 1984, 1986).

There are several instances of higher precipitation: from Alexandersfontein in the northern Cape (Butzer 1984*a,b*), from the Blue Pool tufa on the Gaap Escarpment (Butzer *et al.* 1978*b*; Butzer 1984*a,b*) and from Breek Been Kolk, Hoezar Oost and Kannikwa in the Karoo-Namaqualian Ecozone (Beaumont 1986).

Although the evidence from these sites seems to support the case for wetter conditions during the Last Glacial Maximum north and east of the present-day winter rainfall region, the palaeosols that alternate with dunes in the south-western Cape (Butzer 1984*b*) are not considered to be reliable evidence for higher rainfall at this time. They are dated on the basis of one radiocarbon determination on bone from the A horizon in a palaeosol at Melkbosstrand that Butzer (1984*b*, p. 239) considers indicative of surface stability and a better ground cover rather than a genuinely moist climate. Similarly, the radiocarbon dates of *c.* 18 600 and 18 100 BP on charcoal and ostrich eggshell from the ferruginized horizon at Nelson Bay Cave on the southern Cape coast date the human occupation of the site and not the later period of ferruginization. The lack of change in the incidence of C_3 grasses in the diet of zebras from

Apollo 11 cave at 28° south, which shows that the winter rainfall belt did not shift this far northwards (Vogel 1983), is further evidence against the assumption (Cockcroft *et al.* 1987; Tyson 1986) that winter rainfall was intensified and extended as far as 25° south during the Last Glacial Maximum.

The biological data are not only consistent in indicating a cold and generally dry glacial climate, but also show unusually low species diversity in both plant and animal remains which is a consistent feature of harsh Last Glacial Maximum environments. Again the Boomplaas site with complementary evidence from larger mammals, micromammals, charcoal, and pollen, and its relatively complete sequence, provides the key evidence for the Cape Ecozone (Deacon *et al.* 1984*a*). Together the results demonstrate that an overall temperature reduction of the order of 5°C lower than at present, combined with reduced effective precipitation, is sufficient to impoverish both the vegetation and animal communities of the Cape Ecozone to the extent that they have no modern analogue. Similar impoverishment in vegetation at the Last Glacial Maximum has been noted in the Transvaal at Wonderkrater (Scott 1982*a*), but there are insufficient data from other ecozones to indicate whether this phenomenon was widespread in Southern Africa. More generally, the archaeological proxy data, in highlighting the low incidence of occupation sites of Stone Age people in this time range, confirms that hunter-gatherers in most areas of Southern Africa apparently suffered a marked decrease in population numbers during the Last Glacial (Deacon and Thackeray 1984). The north-western Cape seems to be an exception to this pattern (Beaumont 1986).

7.5 *c.* 16 000–10 000 BP

There is evidence from a wide range of environments for markedly wetter conditions immediately after the Last Glacial Maximum between *c.* 16 000 and 11 000 BP. In the Kalahari and Namib it was wetter at Conception, Meob, Otjimaruru, Gobabis, Urwi Pan, Drotsky's Cave, and the Okavango, but conditions were not as wet as they were between 40 000 and 20 000 BP (Lancaster 1979; Vogel and Visser 1981; Butzer 1984*a,b*; Shaw 1986). In the southern Cape at Boomplaas Cave, on the other hand, the period from *c.* 16 000 to 12 000 was the wettest time of the past 70 000 years. From *c.* 12 000 to 8000 BP conditions around Boomplaas became drier with summer drought (Scholtz 1986). Cave speleothem formation in the Cango caves and in Boomplaas Cave ceased at about 16 000 BP so that no estimate of temperature is available between this time and the middle Holocene (Vogel 1983; Deacon *et al.* 1984*a*). Wetter conditions with summer and winter rainfall increases are indicated by arboreal and shrub pollen at Craigrossie in the Basutolian Ecozone between *c.* 12 600 and 10 700 BP (Scott 1986) and cooler and moister conditions are indicated at Aliwal North between *c.* 13 000 and 12 000 BP (Coetzee 1967) and pre-10 000 BP in East Griqualand (Tusenius 1986). In the winter rainfall region, molerat sizes at Elands Bay cave suggest it was wetter between *c.* 11 000 and 9600 BP, but it was drier before that time (Klein 1984, 1986).

There is clearly some variability for there are data in the interior suggesting a short dry phase between moister ones *c.* 14 000 to 13 000 BP at Alexandersfontein in the Basutolian Ecozone (Butzer *et al.* 1973*a*), and on the Gaap Escarpment in the Kalaharian Ecozone between *c.* 13 500 and 11 500 BP (Butzer *et al.* 1978*b*). Drier conditions persisted at Kathu Pan.

The speed of environmental change can be gauged from the speed with which sea levels rose as glaciers melted at higher latitudes after 16 000 BP in the southern hemisphere. Figure 6.4 illustrates the dated sea levels off the Southern African coast compared against the curve drawn from the Huon Peninsula sea levels, New Guinea, that Chappell and Shackleton (1986) have recalculated after detailed correlation with the oxygen isotope record from deep sea core V19-30. The results are reasonably consistent. Dated land-based sequences in Southern Africa (for example, Boomplaas Cave) confirm that the initial shift in temperatures took place between 16 000 and 14 000 BP, and that this shift was coincident with that observed in the Antarctic (Lorius *et al.* 1979, 1985) and other southern hemisphere continents, and preceded that in the northern hemisphere by between 2000 and 3000 years (Salinger 1984, p. 145; Labeyrie *et al.* 1986, p. 705).

Biological data indicate that marked changes in vegetation and small mammals occurred between 16 000 and 14 000 years ago, but that larger mammal changes in the Cape Ecozone were somewhat slower and post-date 12 000 BP when several of the Late Pleistocene 'giant' bovids became extinct, apparently as the result of the combined effects of environmental changes and more efficient equipment of Later Stone Age hunters (Klein 1980, 1984; Avery 1982*a*; Deacon *et al.* 1984*a*). There is little or no evidence for a comparable change in the larger mammal fauna in most other ecozones where essentially the same taxa were present during the Late Pleistocene and the Holocene.

7.6 *c.* 10 000–0 BP

During the Holocene fluctuations in both temperature and humidity occurred, but the scale of change was considerably lower than that observed between the glacial and interglacial modes of the Late Pleistocene and Holocene. Terms such as 'wetter', 'drier', 'warmer', and 'cooler' should therefore be seen only as relative estimates in the context of the Holocene. Some measure of the scale of change involved can be seen in the temperature estimates calculated from the oxygen isotope content of the Cango speleothem after *c.* 6000 BP (Fig. 6.12) (Vogel and Talma, personal communication) and temperatures calculated from factor analysis of micromammalian samples (Fig. 6.7; Thackeray 1987). They both show that temperature fluctuated less than 1°C around the present day mean, although the Wolkberg speleothem from the Transvaal suggests the variability may have been up to ± 2°C (Talma *et al.* 1974).

7.6.1 *c.* 10 000–7000 BP

Whereas some sites with long and detailed sequences show variability in this time span with evidence for slightly cooler and warmer intervals (Aliwal North, Wonderkrater, Byneskranskop, Boomplaas), the overall picture is one of increasingly warm temperatures.

In the Cape, Transvaalian, Basutolian, and Kalaharian ecozones, where precipitation estimates have been made they indicate a generally drier trend, with the exception of moister conditions indicated by charcoals at Siphiso in Swaziland (Prior and Price Williams 1985) and by a seasonally high water table at Rose Cottage Cave (Butzer 1984*a*). In the Kalaharian Ecozone there is evidence for locally active springs at Etosha between *c.* 11 000 and 9000 BP, and at Stampriet between *c.* 14 000 and 8000 BP (Butzer *et al.* 1978*b*; Heine 1982, 1984). There is no good evidence for markedly wetter conditions raising lake levels to the same extent as has been observed in East Africa in the Early Holocene (Nicholson and Flohn 1980).

In the Cape Ecozone (the only ecozone to show marked differences between glacial and interglacial faunas), larger mammal communities were changing gradually and by 5000 BP were essentially similar to those of historic times. Whereas equids and alcelaphines are still found in small numbers in deposits dating to 10 000 BP (Boomplaas, Nelson Bay Cave, Byneskranskop, Melkhoutboom, Wilton), and vaalribbok, mountain reedbuck, and roan antelope were relatively common in the early Holocene, they are largely absent thereafter indicating that the scrub and bush element in the vegetation increased gradually at the expense of grassland (Klein 1980, 1984). In East Griqualand at Colwinton and Ravenscraig temperatures were warm enough after 10 000 BP to allow alcelaphines and equids to spread to grassland above the escarpment whereas they were uncommon there in the Late Holocene (Opperman

1982, 1984). Charcoal from these sites also shows dry conditions prevailed in the Early Holocene (Tusenius 1986).

To the north and east in the summer rainfall region, pollen analyses from Wonderkrater, Scot, Rietvlei, and Moreletta (Scott 1984*a,b*) indicate relatively dry and cool conditions between *c.* 10 000 and 8500 and somewhat warmer temperatures between *c.* 8500 and 7500 BP.

7.6.2 *c.* 7000–4000 BP

There are few continuous sequences in the first half of the Holocene from which to gauge the scale and timing of temperature changes, but what is available indicates this was the warmest time at Wonderkrater (Scott 1982*a*), Kathu Pan, Equus Cave, and Wonderwerk (Beaumont 1986) in the summer rainfall region, and Groenvlei (Martin 1968) and Boomplaas (Deacon *et al.* 1984*a*) in the all-year rainfall region. There is a break in occupation at Boomplaas between *c.* 9000 and 6400 BP, but the temperature estimates from the micromammals (Thackeray 1987; Fig. 6.7) and the charcoal analyses (Scholtz 1986) show the sample from 6400 BP to have been the warmest in the entire sequence with hot and dry summers suggested by the wood anatomy of the charcoal. The Cango speleothem has no dated samples between *c.* 16 000 and *c.* 5500 BP and although there is a peak in the oxygen isotope 'temperature' curve in the Holocene sequence dating to *c.* 2000 BP, this may not be a true reflection of the scale of temperature change (Vogel and Talma, personal communication). The 'temperatures' derived from Wolkberg Cave (Talma *et al.* 1974) and the Uitenhage aquifer (Heaton *et al.* 1986*a*) show a similar peak dating in the latter half of the Holocene.

From the Transvaalian, Basutolian, and Kalaharian ecozones (Wonderkrater, Alexandersfontein, Wonderwerk, Kathu Pan, Equus Cave, and Okavango) there is some indication of increased moisture between about 7500 and 5000 BP. In the southern Cape moister conditions evident from pollen analyses post-date 7000 BP (Martin 1968), and the stabilization of dunes seen in the Beacon Island soil and interpreted as indicating more humid conditions also probably post-dates 8000–7000 BP (Helgren and Butzer 1977). The warm and moist environments of the mid- to Late Holocene in the southern Cape were marked by an increase in forest taxa in pollen spectra from Groenvlei and Norga. This increase can be related tentatively in the mid-Holocene to a rise in sea level of the order of +1.5 m coincident with warmer temperatures evident from associated warm water diatoms at Groenvlei (Martin 1968) and oyster beds in Langebaan Lagoon (Flemming 1977). Shell at Elands Bay (Yates *et al.* 1986) in a cobble beach 2.8 m above present sea level is dated to nearly 4000 BP.

Micromammals from Byneskranskop in the south-western Cape suggest it was relatively dry between *c.* 6400 and 3900 BP (Avery 1982*a*, p. 338), and drier conditions are reported also from the Basutolian Ecozone at Cornelia, Craigrossie (Scott 1986), Bonawe, Colwinton and Ravenscraig (Tusenius 1986), and from the Transvaalian Ecozone at Siphiso in Swaziland (Prior and Price-Williams 1985) and the Kalaharian Ecozone on the Gaap Escarpment (Butzer *et al.* 1978*b*).

7.6.3 *c.* 4000–0 BP

The last 4000 years have seen low level temperature fluctuations around the present-day mean at most sites after 2000 BP (Vogel 1983; Heaton *et al.* 1986). Between 4000 and 3000 BP wetter intervals have been recorded in lake levels and alluvial deposits in the Kalaharian Ecozone at Makgadikgadi (Helgren 1984), Drotsky's Cave (Grey and Cooke 1977), in the Basutolian Ecozone at Alexandersfontein (Butzer *et al.* 1978*b*) and at Cornelia and Craigrossie (Scott 1986) and possibly Rose Cottage (Butzer 1984*a*). In the Transvaalian Ecozone at Wonderkrater (Scott 1982*a*), Border Cave (Avery 1982*a*), Voigtspost (Horowitz *et al.* 1978) and Moreletta (Scott 1984*b*) it seems to have been somewhat drier. These findings are at variance with those from Siphiso, also in the Transvaalian Ecozone, where the charcoal frequencies are interpreted as indicating relatively moist conditions (Prior and Price-Williams 1985). Charcoals from the East Griqualand sites also suggest relatively moist conditions after 3000 BP, and those from Boomplaas suggest warm and mesic conditions with a lower incidence of drought than occurred in Early and mid-Holocene times (Scholtz 1986).

In the Cape Ecozone somewhat cooler conditions than in the mid-Holocene are suggested by the lower incidence of forest taxa and a higher grass component in pollen spectra in the last 2000 years (Martin 1968; Scholtz 1986), whilst micromammals at Byneskranskop and Elands Bay Cave in the south-western and western Cape suggest it was somewhat moister after 2000 and 3000 BP, respectively, and cooler at Die Kelders during the last 2000 years (Avery 1982*a*, 1983). Optimum conditions for forest expansion in the southern Cape occurred between about 6000 and 2000 BP when it was relatively warm and moist, because thereafter forest taxa were never as common again.

The most important feature of the last 5000–4000 years is the fact that where biological data are available, they show that it was only during this time that modern community alliances were formed. The biogeographic implications of this observation are discussed more fully in the next chapter.

7.7 Summary of radiocarbon dated primary evidence

The main features of the most securely dated primary palaeoclimatic data (equivalent to the level 1 data of Hecht *et al.* 1979) are summarized in Figs 7.1, 7.2, and 7.3. The information is, of course, limited to the last 40 000 years when radiocarbon dates are the most reliable. Caution should be used in the interpretation of these data because the number of dated occurrences is still small and they are geographically scattered. Full details of the dataset are given in Appendix 1.

Figure 7.1a shows that conditions were generally cooler than today prior to 11 000–10 000 BP, but were warmer than today 6000–5000 BP (Fig. 7.1b). A strong warming trend is evident between 10 000 and 8000 BP (Fig. 7.1c) and a cooling trend is shown in the last 4000 years (Fig. 7.1d).

Moisture conditions in the summer and winter rainfall regions are considered separately. Figure 7.2a indicates that throughout the summer rainfall region palaeoclimatic data indicate conditions wetter than the present for much of the last 40 000 years. Clearly, some of this evidence must be considered as 'noise' in the system, and over the last 5000 years at least, it was only slightly wetter than at present. There are two periods with a relatively large number of dates for conditions wetter than at present: 26 000–20 000 BP and 16 000–12 000 BP, with smaller peaks between 7000–6000 BP and 2000–1000 BP. Clear drying trends are evident in the periods 22 000–20 000 BP and 12 000–10 000 BP (Fig. 7.2c). The strongest evidence for a widespread period of conditions drier than the present is for the period 10 000–8000 BP, with a further period of desiccation after 4000 BP (Fig. 7.2b).

In the winter and all-year rainfall region, there are few sites with well dated evidence for moisture changes. The strongest evidence for a period of wetter conditions dates to 14 000–10 000 BP, and again around 4000 BP. It is clear from Fig. 7.3b that conditions were much drier during the period 22 000–16 000 BP.

The bar charts in Figs 7.1 to 7.3 highlight the fact that even primary data in Southern Africa do not provide a clear and unequivocal record for palaeoenvironmental changes, but they do identify the strongest signals. They also show that, although observations are limited, the summer and winter rainfall regions are not consistently out of phase with each other. The relatively crude results from this exercise help to stress the value of palaeoclimatic data based on several different lines of evidence from the same dated sequence. It is from such sequences that we can hope to get the most detailed information on palaeoenvironmental change.

8 Explaining the palaeoenvironmental patterns

Two aspects of palaeoenvironmental research have been explored so far in this book. The first describes the nature, scale, and timing of changes in sedimentological and biological sequences dating to the past 130 000 years in Southern Africa. The second examines the reasons that have been proferred to explain how and why the changes occurred. The results are summarized below, together with remarks on the implications of the observations for biogeography and human population distributions.

8.1 The nature, scale, and timing of change in Late Quaternary palaeoenvironments

All researchers are in agreement that palaeoenvironmental adjustments are directly or indirectly related to the worldwide cycle of glacial and interglacial events of the Late Quaternary. Changing global temperatures were the primary factor in initiating climatic changes that led in turn to adjustments in plant and animal community structure and to disruption of cycles of deposition and erosion. To investigate Southern African climatic and environmental reactions to these fluctuations in temperature we need to be able to interpret accurately the clues that remain in the geological and archaeological record. These interpretations are not always straightforward. By using a combination of inductive and deductive methods we tend to discern patterns, and then try to explain them using modern analogues. To make the explanations believable, it is essential that the interpretations of the field data are critically examined. If this is not done, we may find ourselves explaining events that never happened.

8.1.1 The nature of the terrestrial record

The geological history of Southern Africa in the Quaternary has been characterized by tectonic stability and the main process involved in landform evolution has been erosion rather than deposition. The subcontinent lacks deep sedimentary basins, and alluvial and colluvial mantles are thin. In addition, seasonal climates and soil pH in general are not favourable for the preservation of plant and animal fossils. These conditions make the compilation of a terrestrial record of palaeoenvironmental changes for Southern Africa a complex task.

Early attempts at reconstructing Late Quaternary palaeoenvironments and their relationship to past climatic conditions were largely in the field of the earth sciences. Geological and geomorphological studies continue to be important, but have been complemented by increasing emphasis on palaeontological, palynological, and stable isotope studies. This has enhanced the resolution of reconstructions and has promoted multidisciplinary studies of the same sequences.

The terrestrial record may be patchy and incomplete, but it nevertheless confirms that the scale and timing of climatic changes in Southern Africa were essentially similar to those in other southern hemisphere countries. Many of the details still need resolution. In particular, methods of interpreting past climatic events from alluvial and colluvial processes need refining; more precise data are needed on the distribution patterns of modern plants and animals, and the degree to which they can tolerate changes in environmental parameters. The interaction between animals, plants, and climate is a dynamic one, and we cannot expect that faunal and floral communities will remain static. Faunal remains, pollen, and charcoal data show that plants and animals do not react to climatic change as wholly integrated communities, but each taxon or alliance has its own range of tolerances and adjusts differently. The nature, scale, and timing of changes in Late Quaternary environments as deduced from field and laboratory studies thus serves the dual purpose of informing on the dynamic changes in biogeographic range and ecosystem history, and of providing a test for models of changes in synoptic climates in the subcontinent.

As we have pointed out, the study of sediments, fossils, and their isotopic composition provide proxy rather than direct evidence for past climates. These diverse data have to be interpreted in terms of climatic change as a common denominator and this is done more easily in a relative sense than in absolute terms. For this reason, the most reliable information comes from the longer, more continuous sequences where changes in environmental conditions as gauged from several different lines of evidence at one particular place can be compared against the standard of the present day. By holding place constant, climatic factors can be more easily evaluated.

8.1.2 The scale and timing of the strongest signals

In assessing 'good' evidence from the not-so-good, the most reliable information surely comes from long sequence sites in which several different lines of palaeoclimatic evidence have been studied from the same dated units. In Southern Africa such sites include caves or rock shelters where deposits have been built up through occupation by Stone Age people, carnivores, or owls. The faunal, floral, and sedimentological data provide cross-checks that help in evaluating the signals from various sources (Beaumont *et al.* 1978; H. J. Deacon 1979; Deacon *et al.* 1984*a*; Beaumont 1986). Equally important are the dated long sequence temperature records now available from the Uitenhage aquifer (Heaton *et al.* 1986*a*) and from speleothems (Talma *et al.* 1974; Vogel 1983), as well as offshore sediments and tree rings which have yet to be widely used. Next in importance are regional climatic events that are well dated at a number of different sites, such as the widespread wet phase between *c.* 40 000 and 22 000 BP in the Kalaharian and adjacent ecozones (Lancaster 1979, 1987; Heine 1981, 1982, 1984; Van Zinderen Bakker 1982*a*; Butzer 1984*a,b*), or a series of pollen sequences from different sites in the Transvaal (Scott 1982*a*, 1984*a*). Most lacustrine (pan) alluvial and colluvial deposits represent one or more discrete episodes, and do not provide the same quality of sequential information as cave sites. Their value is therefore determined by the precision of dating and correlation that is possible in a regional framework, and the reliability with which palaeoenvironmental interpretations can be made. Less useful are isolated but dated observations (for example, Horowitz *et al.* 1978), and least useful are undated or poorly dated isolated observations such as faunal remains from open sites (Klein 1980).

By weighting evidence in this way we can usefully select and summarize the strongest signals that have been highlighted in chapters 4–7. The most obvious contrast is that between full interglacial and full glacial temperatures with a clear difference in all ecozones that could have been as much as 9.5°C in the Transvaal and in the Cape was of the order of 5–5.5°C (Talma *et al.* 1974; Vogel 1983; Heaton *et al.* 1986*a*). Effective precipitation was highest during periods when temperatures were intermediate between full interglacial and full glacial, with a longer period of relatively high rainfall during the lead down to the Last Glacial Maximum between about 40 000 and 25 000 BP when temperatures became gradually cooler, and a shorter period in the period immediately following the Last Glacial Maximum between about 16 000 and 12 000 BP when temperatures rose very rapidly (Cooke 1980, 1984; Heine 1982; Deacon *et al.* 1984*a*; Heine and Geyh 1984; Kent and Gribnitz 1985; Teller and Lancaster 1985; Beaumont 1986). The widespread nature of these periods of increased rainfall suggests that evaporation rates, circulation patterns, and relative temperatures of land and sea were at their most favourable for precipitation in Southern Africa during the lead down to and out of a glacial maximum. Where biological evidence is available it shows that the Last Glacial Maximum *sensu stricto* was for the most part exceptionally dry, but in the Kalaharian and adjacent ecozones aridity is suggested only by the absence of evidence for higher precipitation. At other times, regional climatic factors played a stronger role and there is some variability in precipitation in different ecozones and under different rainfall regimes during interglacials, and during periods of gradual temperature reduction after interglacials.

The most reliable evidence for conditions during the Last Interglacial comes from Klasies River (Singer and Wymer 1982; Deacon *et al.* 1986). Oxygen isotope measurements show values close to those of the Present Interglacial at the base of the sequence with a significant regression registered during the time the site was occupied by people making Howiesons Poort-type Middle Stone Age artefacts. The scale and timing of this regression suggest that it correlates with oxygen isotope stage 4 and with the sea level regression between Va and IVa recognized by Chappell and Shackleton (1986) on the Huon Peninsula in New Guinea where a regression is dated to between 80 000 and 60 000 BP. An important implication is that this stage of the Middle Stone Age in other similar sequences, such as at Nelson Bay Cave (Klein 1972*a*), Border Cave (Beaumont *et al.* 1978) and Cave of Hearths (Mason 1962), probably also dates in part to the Last Interglacial and to the terrestrial equivalent of oxygen isotope stage 4.

Although biological evidence suggests there was a step-wise temperature reduction with warm and cooler spells during the time period equivalent to stages 5, 4, and 3 of the oxygen isotope sequence (Beaumont *et al.* 1978; Klein 1980, 1986), these events are not precisely dated in Southern Africa, but can be traced in long sequences such as at Klasies River, Border Cave, and Boomplaas Cave where there was an amelioration of temperatures and wetter conditions at *c.* 60 000 BP post-dating the Howiesons Poort or equivalent substage of the Middle Stone Age.

Dune systems in the Kalahari which indicate stronger winds during a drier climate in that ecozone (Lancaster 1981) are undated, but could relate to the early part of the Last Glacial because they became stabilized and vegetated after *c.* 35 000 BP.

The period of highest rainfall in the summer rainfall region (Kalaharian, Karoo-Namaqualian, and Transvaalian ecozones) was between *c.* 40 000 and 25 000 BP when pan and lake levels were substantially higher (Cooke 1975, 1980; Butzer *et al.* 1978*b*; Lancaster

1979; Heine 1981, 1982; Butzer 1984*a,b*). The Boomplaas sequence in the southern Cape also has evidence for higher rainfall than at present at this time, and Scholtz (1986) tentatively suggests from the wood anatomy of charcoal samples that it may have been mostly winter rain. Unfortunately, there is no well dated evidence from the winter rainfall region of the western and south-western Cape for this time period although micromammals at Die Kelders (Avery 1982*a*) indicate moister conditions that could coincide with part of this time zone.

The coldest period of the Last Glacial Maximum is dated to between *c.* 20 000 and 17 000 BP (Talma *et al.* 1974; Vogel 1983; Heaton *et al.* 1986*a*) and where reliable evidence is available, it is accompanied by dry conditions. Exceptions are the Gaap Escarpment and Alexandersfontein (Butzer *et al.* 1978*b*; Butzer 1984*a,b*) and pans in the north-western Cape (Beaumont 1986). Where available, biological evidence from several ecozones shows conditions to have been exceptionally harsh with low species diversity amongst plant and small mammal communities and in some cases plant communities that have no modern analogue (Avery 1982*a*; Scott 1982*a*, 1984*a*; Deacon *et al.* 1984*a*). Xylem analysis of charcoal from Boomplaas, where today rainfall is at all seasons, but with spring and autumn peaks, suggests dry conditions but with a summer rainfall peak (Scholtz 1986). There is again no unequivocal evidence for this time period from the winter rainfall region although sediments from Elands Bay Cave that probably date between about 20 000 and 12 000 BP suggest it was cold and dry (Butzer 1979). The south-western Cape palaeosols described by Butzer (1984*b*) are neither well developed nor well dated. Sea surface temperatures off the western and eastern coasts of southern Africa were 2–5°C cooler than at present and would certainly have lowered the rate of evaporation (Morley and Hays 1979; Prell *et al.* 1979; Martin 1981; Newell *et al.* 1981; Van Zinderen Bakker 1982*a*).

There is some evidence for moister conditions again after the Last Glacial Maximum in the summer rainfall region in the Kalahari, Namib, and north-western Cape (Heine 1982, 1984; Van Zinderen Bakker 1982*a*; Beaumont 1986), and the Boomplaas data show the highest rainfall in the entire sequence between *c.* 16 000 and 11 000 BP (Deacon *et al.* 1984*a*). After *c.* 10 000 BP the relationship between evaporation and precipitation, and between temperatures on land and oceans changed again as ocean temperatures rose (Vincent 1972). In the period between 12 000 and 9000 BP the dating resolution is not quite fine enough in Southern Africa to find widespread support for the short-term temperature fluctuations such as the Bölling, Allerød, and Older and Younger Dryas recognized in Europe, nor for the two-step deglaciation recognized in some deep-sea and polar ice cores (Ruddiman and

Duplessy 1985), although Coetzee (1967) has recognize stages in the pollen sequence at Aliwal North.

Several long sequence sites with good palaeotemp (Uitenhage aquifer, Cango, and Wolkberg speleothem tional breaks between the end of the Last Glacial Ma Early Holocene, but there is no evidence from other so that temperatures at *c.* 9000 BP were higher than during the mid-Holocene as they apparently were in New Zealand (Salinger 1984) and Australia (Harrison *et al.* 1984). Where details are available, it seems that the highest temperatures in Southern Africa occurred between 7000 and 5000 BP. Holocene temperatures generally fluctuated about 1°C about the present day mean (Martin 1968; Scott 1982*a*, 1984*b*; Vogel 1983; Deacon *et al.* 1984; Heaton *et al.* 1986*a*; Thackeray 1987; Vogel and Talma, personal communication).

Rainfall fluctuations during the Holocene are again of a lower order than has been estimated for changes during the Late Pleistocene. In the Early Holocene, locally moister conditions are reported from the Kalaharian Ecozone (Butzer *et al.* 1978; Heine 1982), from Siphiso rock shelter in Swaziland (Prior and Price-Williams 1985), and from Elands Bay Cave in the winter rainfall region (Klein 1984), but elsewhere a drying trend is indicated, and summer drought is suggested by the wood anatomy of charcoals from Boomplaas (Scholtz 1986) and East Griqualand (Tusenius 1986).

The mid-Holocene record is also patchy with drier conditions indicated from Wonderkrater and Siphiso in the Transvaalian Ecozone (Scott 1982*a*; Prior and Price-Williams 1985), East Griqualand in the Basutolian Ecozone (Tusenius 1986), and from charcoal and micromammals from Boomplaas and Byneskranskop at and after 6400 BP in the Cape Ecozone (Deacon *et al.* 1984; Avery 1982*a*, p. 338). Moister conditions are reported from the Kalaharian Ecozone at Wonderwerk, Gaap Escarpment, and Kathu Pan (Butzer *et al.* 1978*b*; Butzer 1984*a,b*; Beaumont 1986), and from the Cape Ecozone at Groenvlei (Martin 1968).

Somewhat cooler temperatures are recorded during the last 4000 years, but with some warmer temperatures apparent from the Cango speleothem around 2000 BP (Vogel and Talma, personal communication). Transvaalian Ecozone pollen sequences show conditions similar to or drier than those of the present (Scott 1982*a*, 1984*a,b*), although Siphiso and East Griqualand suggest it was relatively moist during this time (Prior and Price-Williams 1985; Tusenius 1986). Cooler and drier conditions are also suggested by a retreat of forest in the southern Cape (Martin 1968; Scholtz 1986), but Boomplaas charcoals of the last 2000 years indicate warm and mesic conditions with a lower incidence of drought than in the Early and mid-Holocene (Deacon *et al.* 1984*a*; Scholtz 1986). In the winter rainfall belt of the western Cape, micromammals suggest it was somewhat moister after 3000 BP at Elands Bay Cave and after 2000 BP at Byneskranskop (Avery 1982*a*, 1983). It is apparent that there does not seem to be a consistent pattern during the Holocene between temperature and rainfall fluctuations in the summer and winter rainfall regions. The implication, as Heine (1984, p. 1751) has remarked, is that the variability of rainfall and temperature responses in Southern Africa to global temperature changes shows that regional thermal and hydrologic changes cannot be used as chronostratigraphic reference points. Where detailed biological data are available, it is only within the last 5000–4000 years that floral and faunal communities assumed their present-day character.

8.2 Biogeographic implications

Biological indicators such as plant and animal fossils provide some of the clearest evidence for the scale and timing of past climatic changes in the Late Quaternary, but whereas the effects of climatic forcing on biological communities is readily discernible, it is less easy to explain the processes that may have been involved in these changes. The reason is simply that the role of environmental determinants on the functioning of ecosystems in Southern Africa is generally not well understood.

Recent phenological studies in the fynbos vegetation of the Cape ecozone (Pierce 1984) show that the seasonal rhythms exhibited within a taxon or between taxa in a community are very variable, precluding simple causal explanations in terms of environmental determinants. On the other hand, environmental conditions are important in constraining distributions as has been shown in a study of the phytogeography of the Poaceae (grasses) in Southern Africa where subfamily regions correspond closely to zones defined on modern climatic parameters (Gibbs Russell 1986).

Any change in the length of the growing season, season of precipitation, soil moisture content, or other ecophysiological factors, would affect plant species differentially and would alter the interaction with animals serving as pollinators, with seed dispersal agents, and with herbivores and in turn their predators. In this way there is considerable potential for dynamic changes in biotic communities over time that can be expected to be reflected in the fossil record. Two examples, Boomplaas Cave in the southern Cape and Wonderkrater in the Transvaal, involved different ecosystems, but show parallel and equivalent scales of Late Quaternary changes. They show the strength of the palaeobiological approach in providing data on ecosystem history, or what Poynton (1986) has termed the historical dimension of ecological biogeography.

At Boomplaas the sequence shows changes in biotic communities in the Cango Valley from *c.* 70 000 BP to the present. During this time, charcoal and pollen analyses show that woodland was replaced by treeless karroid vegetation at the Last Glacial Maximum, and in turn by woodland dominated *Olea* in the terminal Pleistocene, by thicket vegetation in the earlier Holocene, and by modern woodland dominated by *Acacia karroo* in the last 5000 years. Similar changes in habitat are seen in the large and small mammals, and there is supporting evidence for the dating and scale of changes in the sediments and stable isotopes (Deacon *et al.* 1983, 1984*a*). The contrast between the biotic communities of the Last Glacial Maximum and those of the later Holocene is dramatic, and is certainly explicable in gross terms as the result of climatic change. The cold and dry conditions at the Last Glacial Maximum favoured communities that were very different in structure and composition from those found in the valley today, and were different also from communities found today at higher altitudes in the surrounding mountains. They were therefore unique combinations of taxa that correlated with unique climatic conditions.

The Holocene record from Boomplaas is important in showing that modern combinations of plant and animal taxa have developed only within the last 5000 years. The implication is that the climate (hot summers, cool winters, and rainfall all-year with peaks in spring and autumn) that plays a major part in determining the present-day range of vegetation, is also the result of factors that have combined only within the last 5000 years. The carbon-13 content of the Cango speleothem (Fig. 6.13) shows a significant change in vegetation in the Late Holocene from plants dominated by C_3 species that grow under cool conditions in the Late Pleistocene and mid-Holocene, *c.* 5500 BP, to plants dominated by summer-growing C_4 species after *c.* 5000 BP. At Boomplaas and Buffelskloof, and at a number of other sites in the southern and western Cape, there is also a shift in the range of larger mammalian species with modern alliances present only within the last 5000 years (Klein 1978*a,b*, 1980, 1984). The larger mammal data, gathered from sites throughout the Cape Ecozone, confirm that this is not a site specific phenomenon.

In the Cango Valley, the replacement in the later Holocene of thicket vegetation by an *Acacia karroo* woodland alliance of C_4 plants is explicable in terms of a relatively subtle climatic change that includes a hot summer and cool moist winter (Cowling 1983*b*, 1984; Pierce 1984; Pierce and Cowling 1984). A change in precipitation and soil-water balance is also indicated in the Late Holocene by the reinitiation of stalagmite formation in the Cango Caves after *c.* 5500 BP. Furthermore, the presence and prominence of the C_4 grass *Themeda triandra* in the Cango Valley together with *Acacia karroo* indicates an alliance of subtropical generalists (Cowling 1983*a*, p. 141). These are taxa that can utilize biseasonal precipitation and, in the case of *T. triandra*, allow it to displace in part the winter-growing C_3 grasses because it can exploit winter and summer growing seasons equally well (Cowling 1983*b*, p. 125). *Acacia karroo* is a species that has been a major invasive in several biomes in the recent past, and although this may have been promoted by the introduction of domestic stock, charcoal studies from Buffelskloof (Deacon *et al.* 1983) show that the initiation precedes domestic stock and is a product of later Holocene environmental conditions. These conditions promote a winter and summer growing season that is different from the Late Pleistocene and Early Holocene climatic regime under which C_3 plants requiring a cool growing season were favoured (Vogel 1978; Cowling 1983*b*; Vogel and Talma, personal communication).

Cowling (1983*a*) has shown that vegetation communities of the south-eastern Cape are sensitive to climatic change, but that different communities would have reacted in different ways to climatic forcing. By correlating available data on phylogeny, speciation and endemism of plant taxa in this region, he has hypothesized that Cape fynbos taxa are limited by edaphic barriers and the flora was therefore never displaced from its present area of distribution during the Pleistocene. Renosterveld, by contrast, would have fluctuated dramatically depending on prevailing climatic conditions, and subtropical thicket and grassland, and Afromontane forest would have been displaced during colder climatic periods and would have become re-established only during warmer and wetter periods. During drier and colder glacial times, karroid taxa would have invaded areas at present occupied by renosterveld and thicket (Cowling 1983*a*, p. 393). This hypothesis again implies that modern climatic conditions correlate with the *Acacia karroo* and *Themeda triandra* alliance, and that hot summers, cool winters, and all-year, but relatively unpredictable rainfall pertained in the southern Cape only within the last 5000 years. Dynamic reconstructions of vegetation history such as this one will be essential if we are to use palaeontological and palaeobotanical studies in more than a descriptive way to evaluate the complex interrelationships of palaeoclimate and biogeographic factors.

The second example is the Wonderkrater spring deposit in the Transvaalian Ecozone where the pollen sequence has been studied by Scott (1982*a*). As in the Cango Valley, woodland of a different character from that of the present grew around the site prior to the Last Glacial Maximum and was replaced by ericaceous heathland taxa between *c.* 25 000 and 11 000 BP. In the period *c.* 11 000–4000 BP savannah vegetation was dominant and it was only from 4000 BP to the present that the modern bushveld taxa were present. Scott (1982*a*) has interpreted this sequence of changes by relating the palaeocommunities to analogues along modern altitudinal and moisture gradients, and on this basis to infer past climatic conditions (Fig. 5.8), but he has also noted that the palaeocommunities are not exactly the same as the modern analogues. With the knowledge that there are no exact modern analogues for past vegetation communities, it will be advantageous to undertake phyto-chorological studies of the Wonderkrater and surrounding flora before drawing detailed climatic inferences, but it is nevertheless important to note that the modern bushveld alliance is only about 4000 years old and, by implication, so is the present-day rainfall and temperature regime.

Present-day ecosystems are the product of biological processes of evolution, speciation, dispersal, and extinction. The association of plant and animal species in communities is not fixed, but is the product of selection operating over time. The research summarized in this book has shown that climatic changes in the Late Quaternary have been a pervasive factor in determining the composition and nature of modern natural communities, and that modern distributions were achieved only in the last 5000–4000 years. However, during the last 2000 years and particularly since the advent of European colonists, the natural communities have been considerably modified by land use patterns. These agricultural and settlement patterns have lowered the diversity of ecosystems, and have reduced their resilience simulating conditions more similar to the Last Glacial than to the Late Holocene. This loss of diversity through extinction or range restriction of taxa is serious because it implies that as Southern Africa moves into the next glacial, communities will be impoverished to an even greater extent than before. The goals of conservation and management must therefore be directed at maintaining diversity in general, at restoring the natural range of taxa, and at reducing the risk of the loss of endangered species. Whereas processual studies on the functioning of modern ecosystems have received welcome attention in the last decade in Southern Africa, there is an equal need to invest research resources in historical studies.

8.3 Implications for human population distributions

As the data summarized in this book have shown, climatic changes of the Late Quaternary induced significant changes in plant and animal distributions, and in turn these affected human habitats, particularly in the availability of food resources. The distribution and density of human populations was constrained by the ability of people to cope with changes in the natural productivity of their environments because prior to the last 2000 years, agriculture was not practised in Southern Africa.

Africa as a whole is a continent of seasonal rainfall, except for the limited area of the equatorial forest biome, the Guinean Ecozone, and, like Australia, is generally arid and prone to episodic droughts. Persistent droughts lasting many years have been recorded historically, and in regions such as the Sahel and Ethiopia are the cause of famine and stress, as well as of high local extinction rates. Demographic studies among the !Kung San in Botswana (Howell 1979) indicate that fertility is linked to environmental conditions through fat levels and lower birth rates are a response to a reduction in food resources. In this way there is a homoeostatic control on population in accord with natural productivity.

In a study of the archaeological evidence for fluctuations in human population levels during the Late Quaternary, H. J. Deacon (in preparation) suggests that it was Late Pleistocene aridity rather than reduced temperatures that was primarily responsible for lower population levels in Southern Africa. Archaeological visibility (measured by the number of Stone Age occupation sites) was relatively high in Middle Stone Age times during the Last Interglacial, and was high again during the Later Stone Age in the Present Interglacial. In most long sequence sites in all ecozones, discontinuities that indicate significant time breaks separate Middle and Later Stone Age occupation horizons, and Last Glacial occurrences are rare.

Figure 8.1 (from H. J. Deacon, personal communication) summarizes the number of radiocarbon dates that have been published for archaeological deposits in various regions of Africa. They date within the past 40 000 years when the method is applicable and are expressed on a relative scale (RND) for the continent so that they can be compared at the same scale even though the actual number of dates varies from region to region. The assumption is that the radiocarbon dates represent the frequency with which archaeologists have intersected human occupation horizons during excavations and in turn this represents a crude indication of the density of people in the landscape at any particular point in time. In all regions there are relatively few occurrences older than 10 000 years, but the Atlas Region, the Sahara, North-east Africa and, to a lesser extent, East Africa show a significant increase in dated horizons from about 10 000 BP that correlates with a period of markedly higher precipitation and a stronger monsoon system in the Early Holocene (Nicholson and Flohn 1980) evident from higher lake levels (Street and Grove 1976). West Africa, the Miombo Region of Central Africa, and Southern Africa, on the other hand, show a significant increase only in the last 5000 years suggesting they were affected by different climatic factors that brought drier conditions during the Early and mid-Holocene, and higher moisture levels only within the last 5000 years (Martin 1968; Scott 1982*a*, 1986; Butzer 1984*a,b*; Deacon *et al.* 1984*a*;

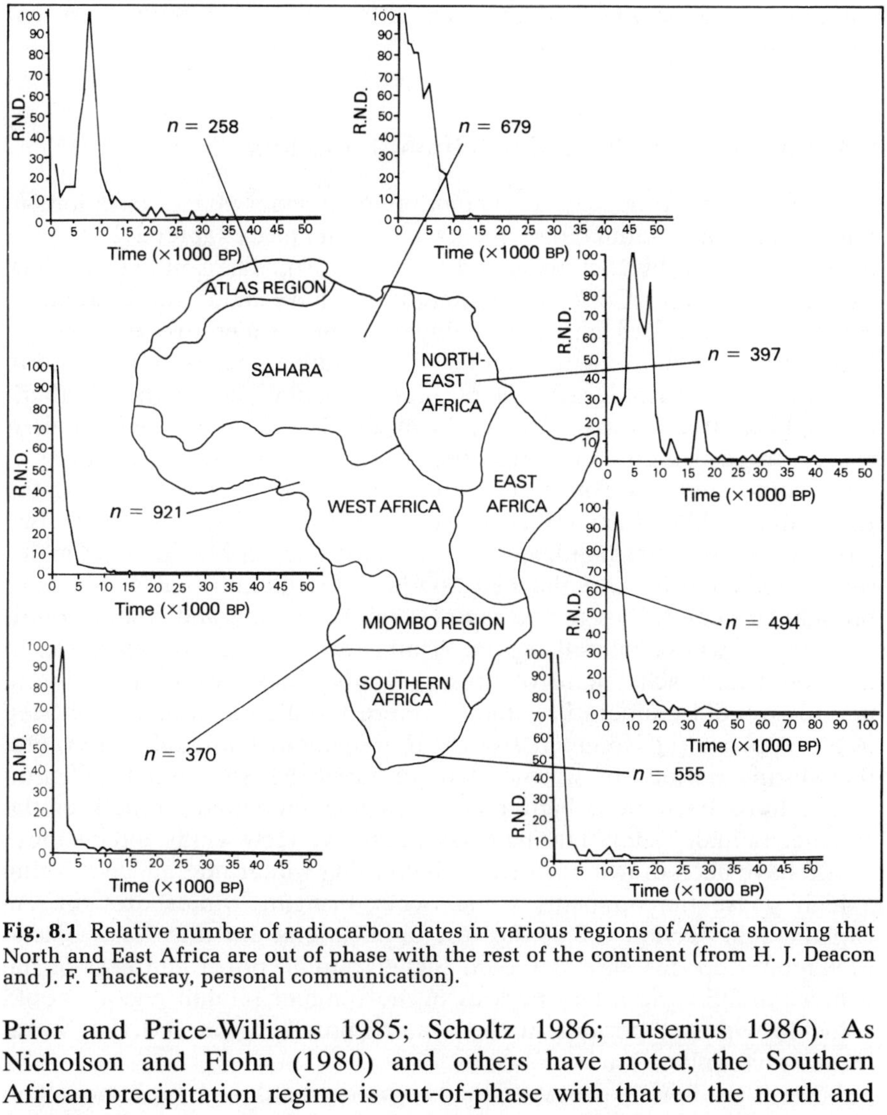

Fig. 8.1 Relative number of radiocarbon dates in various regions of Africa showing that North and East Africa are out of phase with the rest of the continent (from H. J. Deacon and J. F. Thackeray, personal communication).

Prior and Price-Williams 1985; Scholtz 1986; Tusenius 1986). As Nicholson and Flohn (1980) and others have noted, the Southern African precipitation regime is out-of-phase with that to the north and east.

Throughout Africa, the increase in dated archaeological occurrences precedes the appearance of food production by several thousand years suggesting that the spread of food production (which originated earlier in North and Eastern Africa than in West and Southern Africa) was a consequence rather than a cause of Holocene population increases. On the other hand, food production has allowed populations to climb to density levels far beyond those that could have been maintained by hunter-gatherer subsistence systems alone (H. J. Deacon, personal communication).

8.4 Implications for palaeoclimatic models

Climatic models proposed to explain regional changes and variations in rainfall have been refined over the past 25 years (see chapter 2.2) and the most recent ones by Nicholson and Flohn (1980), Harrison *et al.* (1984), Tyson (1986), and Cockcroft *et al.* (1987) have made more sophisticated use of modern climatic analogues. The former two explore the implications of movements of the ITCZ south of its present position because of a much larger extent of continental ice in the northern hemisphere at the Last Glacial Maximum. For Southern Africa their model suggests more vigorous westerlies situated further north than today, but a reduction in precipitation, especially over land during the Last Glacial Maximum (Harrison *et al.* 1984, p. 29). The latter two models, on the other hand, suggest that Last Glacial Maximum climates were dominated by an enhanced Southern Oscillation that would have brought dry conditions to the summer rainfall region and wetter conditions to the winter rainfall region. Winter rains would have extended to 25° south and well inland of present limits. They assume that during periods of cold temperatures the Southern Oscillation would have been in a low phase (i.e. summer drought in the summer rainfall region) and that during periods of warmer temperatures the Southern Oscillation would have been in a high phase bringing increased rainfall to the summer rainfall region. During a low phase, westerly winds and cyclones would have been drawn northwards increasing winter rainfall, and during a high phase the opposite would occur bringing winter drought. As explained in section 2.2.3, the test implications are that Last Glacial Maximum deposits should record higher rainfall north and east of the winter rainfall region. Dry periods in the summer rainfall region should coincide with wet periods in the winter rainfall region and vice versa, should this explanation be correct.

There is a difference between the two models, although both predict an equatorwards movement of the westerlies and winter rains. Harrison *et al.* (1984) suggest Last Glacial Maximum rainfall was probably similar to or lower than at present in the modern winter rainfall area, with Nicholson and Flohn (1980) suggesting increased winter rain to the north of this area (Fig. 2.7). Tyson (1986) and Cockcroft *et al.* (1987), on the other hand, predict higher rainfall from the westerlies both in the present day winter rainfall region, and to the north and east. An important corollary of their argument is the negative correlation between the summer and winter rainfall regions. Neither of these predictions is convincingly met in the data summarized here, mostly because there are so few well dated observations for past climates in the winter rainfall region. However, what little evidence there is (see chapter 6.6.4), supports Harrison *et al.* (1984) in suggesting that conditions in the surrounds of Elands Bay Cave in the western Cape at the Last Glacial Maximum, where deposits are dated within the broad limits of *c.* 20 000 to 11 000 BP, were as cold and dry as they were elsewhere in both summer and all-year rainfall regions (Butzer 1979; Parkington 1986). Although wetter conditions are reported in the north-western Cape and in the Kalaharian Ecozone during the Last Glacial Maximum (Butzer 1984*a,b*; Beaumont 1986), there is no way of testing yet whether this was the result of summer or winter rainfall. The evidence against the latter is the lack of change to C_3 grasses that would be expected at Apollo 11 at 28° south had winter rainfall penetrated to 25° south during the Last Glacial Maximum (Vogel 1983). The only technique that has been developed so far to indicate season of rainfall is xylem analysis used on charcoal samples from Boomplaas and this tentatively suggests dry conditions, but summer rainfall during the Last Glacial Maximum in a region that today receives rain all year round (Scholtz 1986). The fact that the southern Cape mountains have no block streams and that no true eboulis sec is found in caves in the region also suggests that the season of maximum cold did not coincide with a rainfall peak and argues against higher winter rainfall during the Last Glacial Maximum.

We have to conclude that the most reliable palaeoenvironmental evidence available in Southern Africa does not support a model in which the summer and winter rainfall regions were out of phase in the last glacial cycle. On the other hand, this is not to say that the model may not be proved correct when there are more data from the winter rainfall region. At present, we simply do not have enough convincing evidence to support or reject the model outright. As Heine (1984) has shown, combined data from Southern Africa indicate that higher rainfall occurred in the 15 000 years before and in the 4000–5000 years after the Last Glacial Maximum when temperatures were about 2–4°C cooler than at present. During both colder and warmer periods than this, the effect was more variable in different regions of Southern Africa and no clear pattern has emerged yet. More importantly, the historical biogeographic implications of the analyses of biological data indicate that modern communities of plant and animal taxa developed only within the last 5000 years when human population densities in Southern Africa suggest productivity was higher than during the Last Glacial and Early Holocene. This observation in turn implies that modern climatic patterns, including the degree of seasonality that existed between summer and winter temperature, and rainfall regimes, and by the same token the boundaries between winter, summer, and all-year rainfall zones, also date within the last 5000 years. If this is indeed the case, then modern climate analogues may be inappropriate for modelling palaeoclimates for periods older than 5000 years ago in Southern Africa. The challenge for future research will be to test these implications by enlarging the palaeoenvironmental data base and by extending modern studies to provide viable analogues that have greater power to explain past patterns of plant and animal distributions, and the climatic conditions that affect sedimentary sequences. Biogeographic studies, in which environmental changes from a number of different lines of evidence can be integrated, have this potential.

Appendix. List of sites with radiocarbon dated evidence for palaeoclimatic conditions in Southern Africa. Only occurrences that have given reliable primary data are included. The dates are summarized in Figs 7.1, 7.2, and 7.3.

Ecozone	Site	Date	Error	Lab Number	Palaeoclimatic conditions: Temperature	Palaeoclimatic conditions: Moisture	Source of Palaeo-climatic evidence	References
Cape	Boomplaas	32100	700	UW-304	cooler	wetter	pollen	Deacon *et al.* 1984*a,b*;
		21100	420	UW-300	cooler	dry	charcoal	J. Deacon 1984;
		17830	180	Pta-3283		dry	microfauna	Avery 1982*a*
		14200	240	UW-301	warming trend	wet	macrofauna	
		12060	105	Pta-1828	warming trend	wet	,,	
		10425	125	UW-411	warming trend	drying trend	,,	
		9100	135	UW-410	warming trend	drying trend	,,	
		6400	75	UW-306	warm	dry	,,	
		1510	75	UW-307	as at present	as at present	,,	
Cape	Elands Bay Cave	20180	220	Pta-1597		dry	macro and microfauna	Parkington 1986;
		12450	280	Gak-4338		dry	,,	Klein 1980;
		11400	140	UW-192		wetter	,,	Avery 1983
		9600	90	Pta-686		wetter	,,	
Cape	Klasies River Mouth	4755	95	GX-0970		wetter	cave sediments	Butzer 1978;
		4695	180	GX-0973		wetter	,,	Singer and Wymer 1982
		2525	85	GX-0969		wetter	,,	
Cape	Nelson Bay Cave	18660	110	GrN-5884	cold	dry	microfauna	Klein 1972*a,b*;
		18100	550	UW-175	cold	dry	,,	Fairhall *et al.* 1976; Avery 1982*a*
Cape	Loerie	4010	70	UW-169		wetter	alluvial channel fill	Butzer and Helgren 1972
		1330	110	UW-203		wetter	,,	
Cape	Norga	1818	55	Pta-2646		wetter trend	pollen	Scholtz 1986
		2570	45	Pta-2279		wetter trend	,,	
		2980	80	Pta-2683		wetter trend	,,	
Cape	Hangklip	7280	130	GrN-6361	no change	no change	pollen	Scott 1984*a*
		6520	100	GrN-6360	no change		,,	
		1580	50	GrN-6359	no change		,,	
Cape	Groenvlei	6870	160	Y-467		wetter trend	pollen	Martin 1968
		1905	60	Y-466		drying trend	,,	

Appendix—*continued*

Ecozone	Site	Date	Error	Lab Number	Palaeoclimatic conditions:		Source of Palaeo-climatic evidence	References
					Temperature	Moisture		
Cape	Buffelskloof	5120	65	Pta-1981	warmer	wetter	charcoal	Opperman 1978;
		3840	80	UW-341		drying trend	,,	Deacon *et al.* 1983
Cape	Byneskranskop	12730	185	I-7948	cooler		microfauna	Avery 1982*a*, 1983;
		9760	85	Pta-1587	warming trend		,,	Schweitzer and Wilson 1982
		6540	55	Pta-1905	warmer		,,	
		6370	90	Pta-1772	warmer		,,	
		6100	140	UW-409	warmer		,,	
		3900	55	Pta-1571	warmer		,,	
		3400	55	Pta-1569	as at present		,,	
		3220	45	Pta-1631	as at present		,,	
		1880	50	Pta-1865	as at present		,,	
Cape	Uitenhage aquifer	28000		Pta-3148	13.8°C Estimated		Dissolved gasses	Heaton *et al.* 1986*a*
		27800		Pta-3149	14.5°C Recharge		and stable isotopes in	
		22900		Pta-3143	12.6°C Temperature		groundwater	
		21700		Pta-3144	13.1°C		,,	
		21200		Pta-2223	15.3°C		,,	
		20800		Pta-3165	13.9°C		,,	
		18200		Pta-2222	17.5°C		,,	
		15100		Pta-3147	14.6°C		,,	
		15000		Pta-3152	14.5°C		,,	
		8950			20.6°C		,,	
		5590		Pta-3151	23.3°C		,,	
		4920		Pta-2238	21.0°C		,,	
Karoo	Breek Been Kolk	38300	1700	Pta-3958		wet	fresh/brackish	Beaumont 1986;
		29100	590	Pta-4005		wet	water mollusca	Kent and Gribnitz 1985
		24900	290	Pta-4006		wet	,,	
		13930	130	Pta-1760		wet	,,	
Karoo	Hoezar Oost	17900	190	Pta-3894		wet	mollusca	Beaumont 1986

Basutolian	Alexandersfontein	16010	185	SI-1115		wetter	sediments	Butzer *et al.* 1973*a*;
		15400				wetter	,,	Butzer *et al.* 1979;
		15175				wetter	,,	Butzer 1984*a,b*
		14900				wetter	,,	
		14670	270	SI-3563		wetter	organic soil	
		14500				drier	sediments	
		13950				wetter	,,	
		13895				wetter	,,	
		13800				wetter	,,	
		13680				wetter	,,	
		11285				slightly wetter	,,	
		11230				slightly wetter	,,	
		11025				slightly wetter	,,	
		10875				slightly wetter	,,	
		9310				slightly wetter	,,	
		7905				slightly wetter	,,	
		7035				slightly wetter	,,	
		4475	60	SI-2022		slightly wetter	ostrich egg shell	
		4395				slightly wetter	sediments	
		4075	300	SI-2332		slightly wetter	organic spring deposits	
		3805	75	SI-2586		drier	soil carbonate	
		3370				drier	sediments	
		3290	75	SI-2580		drier	ostrich egg shell	
		2520	65	SI-2582		slightly wetter	spring carbonate	
		2225				slightly wetter	sediment	
		2220				slightly wetter	,,	
		1555				wetter	,,	
		1395				wetter	,,	
		805	100	SI-2024		drier	horn core	
Basutolian	Aliwal North	12600	110	GrN-4011	cooler	wetter	pollen	Coetzee 1967
		12200	180	I-2107	cool	wet	,,	
		11650	170	I-2109	warm (but not as warm	dry	,,	
		11250	180	I-2110	warm as present day)	dry	,,	
		9660	150	GrN-4012	warming trend	drier	,,	
Basutolian	Bonawe	8040	100	Pta-2832		as at present	charcoal	Tusenius 1986
		2250	80	Pta-2820		drier	,,	
Basutolian	Ravenscraig (layer 4)	10020	80	Pta-3194		drier	charcoal	Tusenius 1986
Basutolian	Colwinton	6270	40	Pta-2550		drier	charcoal	Tusenius 1986
		1890	45	Pta-2549		wetter trend	,,	

Appendix—*continued*

Ecozone	Site	Date	Error	Lab Number	Palaeoclimatic conditions: Temperature	Moisture	Source of Palaeo-climatic evidence	References
Basutolian	Craigrossie	10700	130	Pta-3389		dry	pollen	Scott 1986
		10600	100	Pta-3682		dry	,,	
		10600	120	Pta-3675		dry	,,	
		10360	100	Pta-3383		dry	,,	
		9460	110	Pta-3686		dry	,,	
		670	60	Pta-3842		wet	,,	
Basutolian	Cornelia	20000	300	Pta-3674	cool	wet	pollen	Scott 1986
		12700	110	Pta-3845	cool	drier	,,	
Basutolian	Sehonghong	32200	770	Pta-785	cold		frost debris,	Carter 1976
		30900	520	Pta-920	cold		carbon isotopes	
Basutolian	Rose Cottage	29430	520	Pta-211	cold		cave sediments	Butzer 1984*a*
		22700	210	Pta-1416B	cool		,,	
		8600	100	Pta-2067		wetter	,,	
		6850	45	GrN-5299		wetter	,,	
		1100	30	GrN-5298		wetter	,,	
Basutolian	Voigtspost	6350	75	Pta-1520	warmer		pollen	Horowitz *et al.* 1978
		1220	50	Pta-1483		drier	,,	
Basutolian	Bushmans Fountain	20500	900	SI-2858		wetter	spring sediments	Butzer *et al.* 1979
		6820	115	SI-2292		wetter	,,	
Transvaal	Rietvlei R1	10300	150	Pta-514	cool		pollen	Scott and Vogel 1983
		9130	140	Pta-3124	cool		,,	
	R2	7750	170	Pta-2991		dry	,,	
		7130	90	Pta-2981	warmer	wetter	,,	
		6970	90	Pta-2982	warmer	wetter	,,	
		6700	90	Pta-2990	warmer	wetter	,,	
		6650	90	Pta-3064	warmer	wetter	,,	
		6580	70	Pta-306	warmer	wetter	,,	
	R4	1290	50	Pta-2978		as at present	,,	

Transvaal	Scot		5070	60	Pta-1388	as at present		pollen	Scott 1982*b*
			4930	70	Pta-1387	as at present		,,	
			3530	60	Pta-1392	cooler		,,	
			2460	50	Pta-1391	cooler		,,	
Transvaal	Moreletta		5220	55	Pta-128	warmer	wetter	pollen	Scott 1984*b*
			440	40	Pta-129	cooler	slightly drier	,,	
Transvaal	Siphiso		8700	120	Pta-3540	warming trend	wetter	charcoal	Prior 1984
			7600	80	Pta-3533	warming trend	wetter	,,	
Transvaal	Wolkberg		29560	600	Pta-174	colder		oxygen isotope analysis of speleothems	Talma *et al.* 1974
			19800	200	Pta-104	colder			
Transvaal	Border Cave		45000	2750–2200	Pta-1190	colder		microfauna and cave sediments	Beaumont 1978; Butzer *et al.* 1978*a*; Avery 1982*b*; Butzer 1984*b*
			37500	1200	Pta-446	colder		,,	
			36800	1000	Pta-422	colder		,,	
			36100	900	Pta-423	colder		,,	
			35700	1100	Pta-424	colder		,,	
			33000	2000	LJ-2892	colder		,,	
			28500	1800	LJ ?	cool		,,	
			13300	150	Pta-721	cool		,,	
Transvaal	Wonderkrater	W1	34400	1900	I-10128	cooler	wetter	pollen	Scott 1982*a*
			24070	260	Pta-2050	cooler	wetter	,,	
			23370	250	Pta-2249	cooler	wetter	,,	
		W2	19000	180	Pta-2114	cooler		,,	
		W4	24340	260	Pta-1066	cooler		,,	
			22040	240	Pta-1062	cooler		,,	
			15040	90	Pta-2232	cooler		,,	
			14180	110	Pta-1710	cooler		,,	
		W5	9640	80	Pta-1709		dry	,,	
		W6a	8390	85	Pta-1708		dry	,,	
		W6b	6330	75	Pta-2311		dry	,,	
		W9/10	1080	50	Pta-1711			,,	
Kalahari	Equus Cave		7480	80	Pta-2495		similar to present	pollen	Beaumont *et al.* 1984
			2390	60	Pta-2452		similar to present	,,	

Appendix—*continued*

Ecozone	Site	Date	Error	Lab Number	Palaeoclimatic conditions: Temperature	Palaeoclimatic conditions: Moisture	Source of Palaeo-climatic evidence	References
Kalahari	Gaap	32700	1190	SI-3379		wetter	tufa deposits	Butzer *et al.* 1978*b*
		26840	520	SI-1647		wetter	,,	
		25900	400	SI-2042		wetter	,,	
		24400	500	SI-2033		wetter	,,	
		22200	300	SI-2041		wetter	,,	
		19950	300	SI-2035		wetter	,,	
		15980	230	SI-1642		wetter	,,	
		15150	165	SI-2043		wetter	,,	
		14010	170	SI-1643		wetter	,,	
		9550	115	SI-1645		wetter	,,	
		9310	105	SI-1641		wetter	,,	
		7715	90	SI-1646		wetter	,,	
		3155	65	SI-2036		wetter	,,	
		2520	70	SI-1644		wetter	,,	
		1375	65	SI-2038		wetter	,,	
		1205	60	SI-2037		wetter	,,	
		1070	65	SI-2044		wetter	,,	
		515	80	SI-2045		wetter	,,	
Kalahari	Wonderwerk	12400	180	Pta-3441	cooler		microfauna, pollen and	Avery 1981;
		9130	90	Pta-2546	cooler	drier	cave sediments	Beaumont *et al.* 1984;
		8000	80	Pta-3366	warming trend	wetter	,,	Butzer 1984*a,b*
		5970	70	Pta-2545	warmer	wetter	,,	Van Zinderen Bakker 1982*b*;
		5180	70	Pta-2544	warmer		,,	
		4890	70	Pta-2797	cooler	drier	,,	
		4240	60	Pta-2541	cooler	drier	,,	
		2910	60	Pta-2543		slightly wetter	,,	
		1210	50	Pta-2779	as at present	as at present	,,	
Kalahari	Kathu Pan	32100	780	Pta-3591		drier	pollen, sediments	Beaumont *et al.* 1984
		27500	530	Pta-3566		drier	,,	
		26930	750	I-13040		drier	,,	
		19800	280	Pta-3586		drier	,,	
		7350	90	Pta-3073		drier	,,	
		6150	70	Pta-3545			,,	
		7350	90	Pta-3073	warmer	wetter	,,	
		4430	60	Pta-3518	warmer	wetter	,,	

		2990	60	Pta-3510		drier	,,	
		1840	50	Pta-3504			,,	
		5980	120	I-13039	warmer	wetter	,,	
		3620	80	I-13036		drier	,,	
		2700	50	Pta-3582		drier	,,	
Kalahari	Klein Awas Pan	23410	980–875	Hv-9885		wetter	stromatolites	Heine 1982
Kalahari	Auob	28030	4900–3200	Hv-9502		wetter	mollusca	Heine 1982
		4480	290	Hv-9499		wetter	,,	
		3440	540	Hv-9500		wetter	,,	
Kalahari	Molopo	19085	1125–985	Hv-9495		wetter	mollusca	Heine 1982
		16625	195	Hv-9503		wetter	,,	
		16425	300	Hv-9483		wetter	,,	
		15580	350	Hv-8368		wetter	,,	
		13075	1250–1030	Hv-9486		wetter	,,	
		12765	475	Hv-9485		wetter	,,	
Kalahari	Aminuis	33500	880	Pta-4157		wetter	spring tufa	Lancaster 1986
		21900	260	Pta-4128		wetter	,,	
Kalahari	Otjimaruru Pan	30400	600	Pta-4153		wetter	spring tufa	Lancaster 1986
		23700	320	Pta-4221		wetter	,,	
		12050	90	Pta-4144		wetter	,,	
Kalahari	Urwi Pan	16300	700	Pta-2150		wetter	stromatolites	Lancaster 1979
		16000	160	Pta-2146		wetter	,,	
		15790	110	Pta-2212		wetter	,,	
		15700	125	Pta-2213		wetter	,,	
Kalahari	Gobabis	17250	190	Pta-2814		wetter	spring tufa	Butzer 1983
		16840	220	Pta-2815		wetter	,,	
		12050	120	Pta-2817		wetter	,,	
		11840	100	Pta-2816		wetter	,,	

Appendix—*continued*

Ecozone	Site	Date	Error	Lab Number	Palaeoclimatic conditions: Temperature	Moisture	Source of Palaeo-climatic evidence	References
Kalahari	Makgadikgadi	30250	520	Hv-8382		wet	mollusca	Heine 1978
		27350	550	Hv-8370		wet	lacustrine carbonate	
		27050	450	Hv-8379		wet	,,	
		25910	1210–1000	Hv-8371		wet	,,	
		24330	270	Hv-8364		wet	,,	
		20990	1100	GaK-4310		wet	,,	Street and Grove 1976
		20835	355	Hv-8365		wet	,,	Heine 1978
		19420	150	GrN-9719	cooler	wetter	peat	Cooke 1984
		19170	660	Hv-8366		wet	lacustrine carbonate	Heine 1978
Kalahari	Gi-Xangwa	31470	1010	SI-4648		wet	lacustrine carbonate	Helgren and Brooks 1983
		23980	590	SI-4090		wet	,,	
		22250	290	SI-4647		wet	,,	
Kalahari	Etosha Pan	11900	120	Pta-3041		wet	lacustrine carbonate	Rust 1984
		9310	90	Pta-3043		wet	,,	
Kalahari	Kwihabe Cave	34000	1700	Wits-306		wetter	cave sinter	Cooke 1984
		29300	1000	Wits-377		wetter	,,	
		16000	200	Wits-166		wetter	,,	
		15200	300	Wits-164		wetter	,,	
		15100	200	Wits-165		wetter	,,	
		14000	200	Wits-168		wetter	,,	
		13600	200	Wits-167		wetter	,,	
		13200	300	Wits-171		wetter	,,	
		13100	200	Wits-170		wetter	,,	
		13000	200	Wits-169		wetter	,,	
		2500	100	Wits-123		wetter	,,	
		2200	100	Wits-122		wetter	,,	
		750	100	Wits-195		wetter	,,	
Namib	Kannikwa	23300	360	Pta-3978	cooler	wet	peat	Beaumont 1986

Namib	Rössing	41500	1280	Hv-11640	wetter	cave sinter	Heine and Geyh 1984
		37000	1700	Hv-11639	wetter	,,	
		36950	1040	Hv-11638	wetter	,,	
		33590	1200	Hv-11637	wetter	,,	
		29830	660	Hv-11636	wetter	,,	
		29700	1360	Hv-9909	wetter	,,	
		26780	360	Hv-11635	wetter	,,	
		26700	540	Hv-9910	wetter	,,	
		26630	500	Hv-11634	wetter	,,	
		26530	910	Hv-9489	wetter	,,	
Namib	Narabeb	26400	280	Pta-3759	wetter	lacustrine carbonates	Teller and Lancaster 1986*b*
		22500	340	Pta-3704	wetter	,,	
		22300	600	Beta-9116	wetter	,,	
		20320	300	Beta-9115	wetter	,,	
Namib	Gobabeb	21500	260	Pta-2652	drying trend	calcified reed casts	Vogel and Visser 1981
		21300	260	Pta-2651	drying trend	,,	
Namib	Khommabes	31900	460	Pta-2588	wetter	calcified reed casts	Vogel and Visser 1981;
		31600	430	Pta-2589	wetter	,,	Teller and Lancaster 1985
		28500	370	Pta-2591	wetter	,,	
		27400	310	Pta-2590	wetter	,,	
		22400	210	Pta-2584	wetter	calcified pedotubules	
		21500	190	Pta-2604	wetter	calcified termite nest	
		20900	230	Pta-1091	wetter	calcified root cast	
Namib	Tsondab	14300	130	Pta-1502	wetter	calcified silt	Vogel and Visser 1981
		13300	90	Pta-1043	wetter	freshwater snails	
		8640	70	Pta-1501	wetter	calcareous silt	
Namib	Koichab	23800	340	Pta-3184	wetter	calcified reed beds	Lancaster 1984*b*
		23300	330	Pta-3185	wetter	,,	
Namib	Conception-Meob	12550	120	Pta-1647	drying trend	calcified reed stalks	Vogel and Visser 1981
		10900	100	Pta-1238	drying trend	,,	
		11800	90	Pta-1830	drying trend	,,	

References

Acocks, J. P. H. (1953). Veld types of South Africa. *Botanical Survey of South Africa Memoirs* **28**, 1–192.

Aharon, P. (1983). 140,000-yr Isotope climatic record from raised coral reefs in New Guinea. *Nature* **304**, 720–3.

Alexandre, J. (1962). Phenome periglaciaires dans le Basutoland et le Drakensberg du Natal, *Biuletyn Peryglacjaln* **11**, 11–13.

Ambrose, S. and De Niro, M. J. (1986). Reconstruction of African human diet using bone collagen carbon and nitrogen isotope ratios. *Nature* **318**, 321–4.

Armstrong, A. L. and Jones, N. (1936). The antiquity of man in Rhodesia as demonstrated by stone implements of the ancient Zambezi gravels south of the Victoria Falls. *Journal of the Royal Anthropological Institute* **66**, 331.

Ash, J. E. and Wasson, R. J. (1983). Vegetation and sand movement in the Australian desert dunefield. *Zeitschrift für Geomorphologie* (Suppl.) **45**, 7–25.

Avery, D. M. (1981). Holocene micromammalian faunas from the northern Cape Province, South Africa. *South African Journal of Science* **77**, 265–73.

Avery, D. M. (1982*a*). Micromammals as palaeoenvironmental indicators and an interpretation of the late Quaternary in the southern Cape Province, South Africa. *Annals of the South African Museum* **85**, 183–374.

Avery, D. M. (1982*b*). The micromammalian fauna from Border Cave, KwaZulu, South Africa. *Journal of Archaeological Science* **9**, 187–204.

Avery, D. M. (1983). Palaeoenvironmental implications of the small Quaternary mammals of the fynbos region. In *Fynbos palaeoecology: a preliminary synthesis*, South African National Scientific Progress Report. No. 75 (ed. H. J. Deacon, Q. B. Hendey, and J. J. N. Lambrechts), pp. 139–55.

Avery, D. M. (1987). Late Pleistocene coastal environments of the southern Cape Province of South Africa: micromammals from Klasies River Mouth. *Journal of Archaeological Science* **14**, 405–21.

Backman, J., Shackleton, N. J., and Tauxe, L. (1983). Quantitative nannofossil correlation to open ocean deep-sea sections from Plio-Pleistocene boundary at Vrica, Italy. *Nature* **304**, 156–8.

Baillieul, T. A. (1975). A reconnaissance survey of the cover sands in the Republic of Botswana. *Journal of Sedimentological Petrology* **45**, 494–503.

Barwis, J. H. and Tankard, A. J. (1983). Pleistocene shoreline deposition and sea-level history of Swartklip, South Africa. *Journal of Sedimentary Petrology* **53**, 1281–94.

Beaumont, P. B. (1978). Border Cave. Unpublished M.A. thesis, University of Cape Town.

Beaumont, P. B. (1986). Where did all the young men go during 0–18 stage 2? *Palaeoecology of Africa* **17**, 79–86.

Beaumont, P. B. and Vogel, J. C. (1972). On a new radiocarbon chronology for Africa south of the Equator. *African Studies* **31**, 66–89, 155–82.

Beaumont, P. B., De Villiers, H., and Vogel, J. C. (1978). Modern man in sub-Saharan Africa prior to 49 000 years B.P.: a review and evaluation with particular reference to Border Cave. *South African Journal of Science* **74**, 409–19.

Beaumont, P. B., Van Zinderen Bakker, E. M., and Vogel, J. C. (1984). Environmental changes since 32 000 BP at Kathu Pan, northern Cape. In *Late Cainozoic Palaeoclimates of the Southern Hemisphere* (ed. J. C. Vogel), pp. 329–38. Balkema, Rotterdam.

Bernard, E. A. (1962). Interpretation astronomique des pluviaux et interpluviaux du Quaternaire Africain. In *Actes du IVe Congres Panafricain de Préhistoire et de l'étude du Quaternaire* (ed. G. Mortelmans and J. Nenquin), pp. 67–96. Musée Royal de l'Afrique Central Annales-Sciences Humaines 40, Tervuren.

Birch, G. F. (1976). Surficial sediments of Saldanha Bay and Langebaan Lagoon. *Transactions of the Geological Society of South Africa* **79**, 293–300.

Birch, G. F., Du Plessis, A., and Willis, J. P. (1978). Offshore and on land geological and geophysical investigations in the Wilderness Lakes region. *Transactions of the Geological Society of South Africa* **81**, 339–52.

Birkeland, P. W. (1974). *Pedology, Weathering and Geomorphological Research.* Oxford University Press, Oxford.

Bond, G. (1967). River valley stratigraphy, morphology and palaeoclimatology in southern Africa. In *Background to Evolution in Africa* (ed. W. W. Bishop and J. D. Clark) pp. 303–12. University of Chicago Press, Chicago.

Boocock, C. and Van Straten, D. J. (1962). Notes on the geology and hydrogeology of the central Kalahari, Bechuanaland Protectorate. *Transactions of the Geological Society of South Africa* **65**, 125–71.

Boucher, K. (1975). *Global Climate.* English University Press, London.

Bowler, J. M. (1971). Pleistocene salinities and climatic change: evidence from lakes and lunettes in southeastern Australia. In *Aboriginal Man and Environment in Australia* (ed. D. J. Mulvaney and J. Golson), pp. 47–65. Australian National University Press, Canberra.

Bowler, J. M. (1976). Aridity in Australia: age, origins and expression in aeolian landforms and sediments. *Earth Sciences Review* **12**, 279–310.

Bowler, J. M. and Wasson, R. J. (1984). Glacial age environments of inland Australia. In *Late Cainozoic Palaeoclimates of the Southern Hemisphere* (ed. J. C. Vogel), pp. 183–208. Balkema, Rotterdam.

Brain, C. K. (1958). The Transvaal ape-man-bearing cave deposits. *Transvaal Museum Memoirs* **11**, 1–125.

Brain, C. K. (1967). Procedures and some results in the study of Quaternary cave fillings. In *Background to Evolution in Africa* (ed. W. W. Bishop and J. D. Clark), pp. 285–301. University of Chicago Press, Chicago.

Brain, C. K. (1969). New evidence for climatic change during Middle and Late Stone Age times in Rhodesia. *South African Archaeological Bulletin* **24**, 127–43.

Brain, C. K. (1974). The use of microfaunal remains as habitat indicators in the Namib. *South African Archaeological Society, Goodwin Series* **2**, 55–60.

Brain, C. K. and Brain, V. (1977). Microfaunal remains from Mirabib: some evidence of palaeoecological changes in the Namib. *Madoqua* **10**, 285–93.

Brain, C. K. and Meester, J. (1964). Past climatic changes as biological isolating mechanisms in southern Africa. In *Ecological Studies in Southern Africa* (ed. D. H. S. Davis) pp. 332–40. Junk, The Hague.

Brink, J. and Deacon, H. J. (1982). A study of a last interglacial shell midden and bone accumulation at Herolds Bay, Cape Province, South Africa. *Palaeoecology of Africa* **15**, 31–9.

Brook, G. A. (1982). Stratigraphic evidence of Quaternary climatic change at Echo Cave, Transvaal, and a palaeoclimatic record for Botswana and northeastern South Africa. *Catena* **9**, 343–51.

Brown, A. J. V. and Verhagen, B. Th. (1985). Two *Antidorcas bondi* individuals from the Late Stone Age site of Kruger Cave 35/83, Olifantsnek, Rustenburg District, South Africa. *South African Journal of Science* 81–102.

Buckley, R. L. (1981). Parallel dunefield ecosystems: southern Kalahari and central Australia. *Journal of Arid Environments* **4**, 287–98.

Butzer, K. W. (1971). Fine alluvial fills in the Orange and Vaal basins of South Africa. *Proceedings of the Association of American Geographers* **3**, 41–8.

Butzer, K. W. (1973*a*). Geology of Nelson Bay Cave, Robberg, South Africa. *South African Archaeological Bulletin* **28**, 87–110.

Butzer, K. W. (1973*b*). Pleistocene 'periglacial' phenomena in southern Africa. *Boreas* **2**, 1–11.

Butzer, K. W. (1978). Sediment stratigraphy of the Middle Stone Age sequence at Klasies River Mouth, Tsitsikamma coast, South Africa. *South African Archaeological Bulletin* **33**, 141–51.

Butzer, K. W. (1979). Geomorphology and geo-archaeology at Elandsbaai, western Cape, South Africa. *Catena* **6**, 157–66.

Butzer, K. W. (1983). Global sea level stratigraphy: an appraisal. *Quaternary Science Review* **2**, 1–15.

Butzer, K. W. (1984*a*). Late Quaternary palaeoenvironments in South Africa. In *Southern Africa palaeoenvironments and Prehistory* (ed. R. G. Klein), pp. 1–64. Balkema, Rotterdam.

Butzer, K. W. (1984*b*). Late Quaternary environments in South Africa. In *Late Cainozoic Palaeoclimates of the Southern Hemisphere* (ed. J. C. Vogel), pp. 235–64. Balkema, Rotterdam.

Butzer, K. W. and Helgren, D. M. (1972). Late Cenozoic evolution of the Cape coast between Knysna and Cape St Francis, South Africa. *Quaternary Research* **2**, 143–69.

Butzer, K. W., Isaac, G. L., Richardson, J. L., and Washbourn-Kamau, C. (1972). Radiocarbon dating of East African lake levels. *Science* **175**, 1069–76.

Butzer, K. W., Fock, G. J., Stuckenrath, R., and Zilch, A. (1973*a*). Paleohydrology of Late Pleistocene lake Alexandersfontein, Kimberley, South Africa. *Nature* **243**, 328–30.

Butzer, K. W., Helgren, D. M., Fock, G. J., and Stuckenrath, R. (1973*b*). Alluvial terraces of the lower Vaal River, South Africa, a reappraisal and reinvestigation. *Journal of Geology* **81**, 341–62.

Butzer, K. W., Beaumont, P. B., and Vogel, J. C. (1978*a*). Lithostratigraphy of Border Cave, KwaZulu, South Africa, a Middle Stone Age sequence beginning *c.* 195 000 B.P. *Journal of Archaeological Science* **5**, 317–41.

Butzer, K. W., Stuckenrath, R., Bruzewicz, A. J., and Helgren, D. M. (1978*b*). Late Cenozoic paleoclimates of the Gaap Escarpment, Kalahari margin, South Africa. *Quaternary Research* **10**, 310–39.

Butzer, K. W., Fock, G. J., Scott, L., Stuckenrath, R. (1979). Dating and context of rock engravings in southern Africa. *Science* **203**, 1201–14.

Cable, J. H. C., Scott, K., and Carter, P. C. (1980). Excavations at Good Hope shelter, Underberg district, Natal. *Annals of the Natal Museum* **24**, 1–34.

Caratini, C. and Tissot, C. (1982). Palynological study of Pleistocene sediment cores from Walvis Ridge. *Palaeoecology of Africa* **15**, 227.

Carter, P. L. (1976). The effects of climatic change on settlement in eastern Lesotho during the Middle and Late Stone Age. *World Archaeology* **8**, 197–206.

Cartmill, M. (1967). The early Pleistocene mammalian fauna of sub-Saharan Africa and their ecological significance. *Quaternaria* **9**, 169–97.

Chappell, J. and Shackleton, N. J. (1986). Oxygen isotopes and sea level. *Nature* **324**, 137–40.

Clark, J. D. (1950). *The Stone Age Cultures of Northern Rhodesia.* South African Archaeological Society, Claremont.

Clark, J. D. (1957). A review of prehistoric research in Northern Rhodesia and Nyasaland. In *Proceedings of the Third Pan-African Congress on Prehistory, Livingstone 1955* (ed. J. D. Clark), pp. 412–32. Chatto & Windus, London.

CLIMAP Project Members (1976). The surface of the Ice-Age earth. *Science* **191**, 1131–7.

CLIMAP Project Members (1984). The Last Interglacial ocean. *Quaternary Research* **21**, 123–224.

Cockcroft, M. J., Wilkinson, M. J., and Tyson, P. D. (1987). The application of a present-day climatic model to the Late Quaternary in southern Africa. *Climatic Change* **10**, 161–91.

Coetzee, J. A. (1967). Pollen analytical studies in East and Southern Africa. *Palaeoecology of Africa* **3**, 1–146.

Coetzee, J. A. (1978). Climatic and biological changes in south-western Africa during the Late Cainozoic. *Palaeoecology of Africa* **10**, 13–29.

Cooke, H. B. S. (1957). The problem of Quaternary glacio-pluvial correlation in East and southern Africa. In *Proceedings of the Third Pan-African Congress on Prehistory, Livingstone 1955* (ed. J. D. Clark), pp. 51–5. Chatto & Windus, London.

Cooke, H. B. S. (1962). The Pleistocene environment in southern Africa: hypothetical vegetation in southern Africa during the Pleistocene. *Annals of the Cape Provincial Museums* **2**, 11–15.

Cooke, H. B. S. (1964). The Pleistocene environment in southern Africa. In *Ecological Studies in Southern Africa* (ed. D. H. S. Davis), pp. 1–23. Junk, The Hague.

Cooke, H. B. S. (1967). The Pleistocene sequence in South Africa and problems of correlation. In *Background to Evolution in Africa* (ed. W. W. Bishop and J. D. Clark), pp. 175–84. University of Chicago Press, Chicago.

Cooke, H. J. (1975). The paleoclimatic significance of caves and adjacent landforms in western Ngamiland, Botswana. *Geographical Journal* **141**, 430–44.

Cooke, H. J. (1980). Landform evolution in the context of climatic change and neo-tectonism in the middle Kalahari of north central Botswana. *Transactions of the Institute of British Geographers (N.S.)* **5**, 80–99.

Cooke, H. J. (1984). The evidence from northern Botswana of Late Quaternary climatic change. In *Late Cainozoic Palaeoclimates of the Southern Hemisphere* (ed. J. C. Vogel), pp. 265–78. Balkema, Rotterdam.

Cooke, H. J. and Verstappen, H. T. (1984). The landforms of northern Botswana, with a consideration of the chronology of the evolution of Lake palaeo-Makgadikgadi. *Zeitschrift für Geomorphologie* **28**, 1–19.

Corvinus, G. (1983). *The Raised Beaches of CDM on the West Coast of South West Africa/Namibia.* Verlag C. H. Beck, Munich.

Cowling, R. M. (1983*a*). Phytochorology and vegetation history in the south-eastern Cape, South Africa. *Journal of Biogeography* **10**, 393–419.

Cowling, R. M. (1983*b*). The occurrence of C_3 and C_4 grasses in fynbos and allied shrublands in the south eastern Cape, South Africa. *Oecologia (Berlin)* **58**, 121–7.

Cowling, R. M. (1984). A syntaxonomic and synecological study in the Humansdorp region of the Fynbos Biome. *Bothalia* **15**, 175–227.

Cronin, T. M., Szabo, B. J., Ager, T. A., Hazel, J. E., and Owens, J. P. (1981). Quaternary climates and sea levels of the U.S. Atlantic coastal plain. *Science* **211**, 233–40.

Cruz-Uribe, K. (1983). The mammalian fauna from Redcliff cave, Zimbabwe. *South African Archaeological Bulletin* **38**, 7–16.

Curtis, B. A., Tyson, P. D., and Dyer, T. G. J. (1979). Dendrochronological age determination of *Podocarpus falcatus. Trees in South Africa* **31**, 2–7.

Dardis, G. F. and Granger, J. E. (1986). Contemporary periglacial phenomena in the Natal Drakensberg, South Africa. *Palaeoecology of Africa* **17**, 89–94.

Dart, R. A. (1925). *Australopithecus africanus:* the ape-man of South Africa. *Nature* **115**, 195–9.

Davies, O. (1976). The older coastal dunes of Natal and Zululand and their relation to former shorelines. *Annals of the South African Museum* **71**, 19–32.

Davies, O. (1980). Late Interglacial shorelines in the south Cape. *Palaeontologia Africana* **23**, 153–71.

Deacon, H. J. (1964). Report on the excavations at Elandsfontein, January 1964. Unpublished MS.

Deacon, H. J. (1970). The Acheulian occupation at Amanzi Springs, Uitenhage district, Cape Province. *Annals of the Cape Provincial Museums* **8**, 89–189.

Deacon, H. J. (1979). Excavations at Boomplaas Cave—a sequence through the Upper Pleistocene and Holocene in South Africa. *World Archaeology* **10**, 241–57.

Deacon, H. J. (1983). Another look at the Pleistocene climates of South Africa. *South African Journal of Science* **79**, 325–8.

Deacon, H. J. and Deacon, J. (1986). Late Pleistocene and Holocene climates and human responses. *South African Journal of Science* **82**, 74–5.

Deacon, H. J. and Thackeray, J. F. (1984). Late Quaternary environmental changes and implications from the archaeological record in southern Africa. In *Late Cainozoic Palaeoclimates of the Southern Hemisphere* (ed. J. C. Vogel), pp. 375–90. Balkema, Rotterdam.

Deacon, H. J., Scholtz, A., and Daitz, L. D. (1983). Fossil charcoals as a source of palaeoecological information in the fynbos region. In *Fynbos Palaeoecology: a Preliminary Synthesis.* South African National Scientific Progress Report 75 (ed. H. J. Deacon, Q. B. Hendey, and J. J. N. Lambrechts), pp. 174–82. C.S.I.R., Pretoria.

Deacon, H. J., Deacon, J., Scholtz, A., Thackeray, J. F., Brink, J. S. and Vogel, J. C. (1984*a*). Correlation of palaeoenvironmental data from the Late Pleistocene and Holocene deposits at Boomplaas cave, southern Cape. In *Late Cainozoic Palaeoclimates of the Southern Hemisphere* (ed. J. C. Vogel), pp. 339–52. Balkema, Rotterdam.

Deacon, H. J., Geleijnse, V. B., Thackeray, A. I., Thackeray, J. F., and Tusenius, M. L. (1986). Late Pleistocene cave deposits in the southern Cape: current research at Klasies River. *Palaeoecology of Africa* **17**, 31–8.

Deacon, J. (1974). Patterning in the radiocarbon dates for the Wilton/Smithfield complex in southern Africa. *South African Archaeological Bulletin* **29**, 3–18.

Deacon, J. (1978). Changing patterns in the late Pleistocene/early Holocene prehistory of southern Africa as seen from the Nelson Bay Cave stone artifact sequence. *Quaternary Research* **10**, 84–111.

Deacon, J. (1984). Later Stone Age people and their descendants in southern Africa. In *Southern African Palaeoenvironments and Prehistory* (ed. R. G. Klein), pp. 221–328. Balkema, Rotterdam.

Deacon, J., Lancaster, N., and Scott, L. (1984*b*). Evidence for Late Quaternary climatic change in southern Africa: summary of the proceedings of the SASQUA workshop held in Johannesburg, September 1983. In *Late Cainozoic Palaeoclimates of the Southern Hemisphere* (ed. J. C. Vogel), pp. 391–404. Balkema, Rotterdam.

De Graaff, G. (1960). A preliminary investigation of the mammalian microfauna in Pleistocene deposits of caves in the Transvaal System. *Palaeontologia Africana* **7**, 59–118.

Dingle, R. V. and Rogers, J. (1972*a*). Pleistocene palaeogeography of the Agulhas Bank. *Transactions of the Royal Society of South Africa* **40**, 155–65.

Dingle, R. V. and Rogers, J. (1972*b*). Effects of sea-level changes on the Pleistocene palaeoecology of the Agulhas Bank. *Palaeoecology of Africa* **6**, 55–8.

Dunwiddie, P. W. and La Marche, V. C. (1980). A climatically responsive tree-ring record from *Widdringtonia cedarbergensis,* Cape Province, South Africa. *Nature* **286**, 796–7.

Du Toit, A. L. (1926). The Mier country. *South African Geographical Journal* **9**, 21–6.

Du Toit, A. L. (1947). Palaeolithic environments in Kenya and the Union—a contrast. *South African Archaeological Bulletin* **2**, 28–40.

Du Toit, A. L. (1954). *The Geology of South Africa*, 3rd edn. Oliver & Boyd, London.

Dyer, T. G. J. (1976). Meridional interactions between rainfall and surface pressure. *Nature* **264**, 48–9.

Dyer, T. G. and Marker, M. E. (1979). On some aspects of Lesotho hollows. *Zeitschrift für Geomorphologie* **23**, 256–70.

Ebert, J. I. and Hitchcock, R. K. (1978). Ancient lake Makgadikgadi, Botswana, mapping, measurement and palaeoclimatic significance. *Palaeoecology of Africa* **10**, 47–56.

Einsele, G., Herm, D., and Schwarz, U. (1977). Variations du niveau de la mer sur plateforme continentale et la côte Mauritanienne vers la fin de la glaciation de Wurm et a l'Holocène. *Bulletin d'Association Sénégalaise d'Etude du Quaternaire* **51**, 35–49.

Ellenberger, P. (1960). Le Quaternaire au Basutoland. *Bulletin de Société Prehistoire Française* **57**, 439–75.

Fairbridge, R. W. (1983). The Pleistocene–Holocene boundary. *Quaternary Science Review* **1**, 215–44.

Fairhall, A. W., Young, A. W., and Erikson, J. L. (1976). University of Washington dates IV. *Radiocarbon* **18**, 221–39.

Fitzpatrick, R. W. (1978). Periglacial soils with fossil permafrost horizons. *Annals of the Natal Museum* **23**, 475–84.

Flemming, B. W. (1976). Distribution of recent sediments in Saldanha Bay and Langebaan Lagoon. *Transactions of the Royal Society of South Africa* **42**, 317–40.

Flemming, B. W. (1977). Langebaan Lagoon: a mixed carbonate-siliclastic environment in a semi-arid climate. *Sedimentary Geology* **18**, 61–96.

Flint, R. F. (1959*a*). On the basis of Pleistocene correlation in East Africa. *Geological Magazine* **96**, 265–84.

Flint, R. F. (1959*b*). Pleistocene climates in eastern and southern Africa. *Bulletin of the Geological Society of America* **70**, 343–74.

Flint, R. F. and Bond, G. (1968). Pleistocene sand ridges and pans in western Rhodesia. *Bulletin of the Geological Society of America* **79**, 299–314.

Flohn, H. (1967). Bemerkungen zur Asymmetrie der atmospharischen Zirkulation. *Annalen der Meteorologie, Neue Folge* **3**, 76–80.

Flohn, H. (1984). Climate evolution in the southern hemisphere and the equatorial region during the Late Cenozoic. In *Late Cainozoic Palaeoclimates of the Southern Hemisphere* (ed. J. C. Vogel), pp. 5–20. Balkema, Rotterdam.

Fryberger, S. G. (1980). Dune forms and wind regime, Mauretania, West Africa: implications for past climate. *Palaeoecology of Africa* **12**, 79–96.

Gasse, F. and Street, F. A. (1978). Late Quaternary lake level fluctuations and environments of the northern Rift Valley and Afar region (Ethiopia and Djibouti). *Palaeogeography, Palaeoclimatology and Palaeoecology* **24**, 279–325.

Gates, W. L. (1976). Modeling the Ice-Age climate. *Science* **191**, 1138–44.

Gevers, T. W. (1936). The morphology of western Damaraland and the adjoining

Namib Desert. *South African Geographical Journal* **19**, 61–79.

Gibbs Russell, G. E. (1986). Significance of different centres of diversity in subfamilies of Poaceae in southern Africa. *Palaeoecology of Africa* **17**, 183–92.

Gillooly, J. F. (1975). A preliminary assessment of the potential offered by some indigenous tree species in the Rustenburg district, for dendrochronology in South Africa. Unpublished Honours seminar paper. University of the Witwatersrand Department of Geography and Environmental Studies, Johannesburg.

Giresse, P. and Kouyoumontzakis, G. (1974). Observations sur le Quaternaire cotier et sous-marin du Congo et des regions limitrophes: aspects eustatiques et climatiques. *Bulletin d'Association Sénéglaise d'Etude Quaternaire* **42** and **43**, 45–61.

Gooch, W. D. (1881). The Stone Age of South Africa. *Journal of the Royal Anthropological Institute* **11**, 124–82.

Goodwin, A. J. H. and Van Riet Lowe, C. (1929). The Stone Age cultures of South Africa. *Annals of the South African Museum* **27**, 1–289.

Goudie, A. S. (1973). *Duricrusts in Tropical and Subtropical Landscapes.* Clarendon Press, Oxford.

Goudie, A. S. and Thomas, D. S. G. (1985). Pans in southern Africa, with particular reference to South Africa and Zimbabwe. *Zeitschrift für Geomorphologie NF* **29**, 1–19.

Goudie, A. S., Allchin, B., and Hegde, K. T. M. (1973). The former extensions of the great Indian sand desert. *Geographical Journal* **139**, 243–57.

Gregory, J. W. (1921). *The Rift Valleys and Geology of East Africa.* Seeley, Service & Co., London.

Grey, D. R. C. and Cooke, H. J. (1977). Some problems in the Quaternary evolution of the landforms of northern Botswana. *Catena* **4**, 123–33.

Grindley, J. R. (1986). Pleistocene rains in the north-western Karoo. *South African Journal of Science* **82**, 178–9.

Grine, F. E. and Klein, R. G. (1985). Pleistocene and Holocene human remains from Equus Cave, South Africa. *Anthropology* **8**, 55–98.

Grove, A. T. (1969). Landforms and climatic change in the Kalahari and Ngamiland. *Geographical Journal* **135**, 191–212.

Grove, A. T. and Warren, A. (1968). Quaternary landforms and climate on the south side of the Sahara. *Geographical Journal* **134**, 194–208.

Hall, M. (1976). Dendroclimatology, rainfall and human adaptation in the Later Iron Age of Natal and Zululand. *Annals of the Natal Museum* **22**, 693–703.

Harper, G. (1969). Periglacial evidence in southern Africa during the Pleistocene. *Palaeoecology of Africa* **4**, 71–101.

Harrison, M. S. J. (1986). A synoptic climatology of South African rainfall variability. Unpublished PhD. thesis, University of the Witwatersrand, Johannesburg.

Harrison, S. P., Metcalfe, S. E., Street-Perrott, F. A., Pittock, A. B., Roberts, C. N., and Salinger, M. J. (1984). A climate model of the Last Glacial/Interglacial transition based on palaeotemperature and palaeohydrological evidence. In *Late Cainozoic Palaeoclimates of the Southern Hemisphere* (ed. J. C. Vogel), pp. 21–34. Balkema, Rotterdam.

Hastenrath, S. and Wilkinson, J. (1972). A contribution to the periglacial morphology of Lesotho, southern Africa. *Biuletyn Periglacjalny* **22**, 151–67.

Heath, G. R. (1979). Simulations of a glacial palaeoclimate by three different atmospheric general circulation models. *Palaeogeography, Palaeoclimatology, and Palaeoecology* **26**, 291–303.

Heaton, T. H. E. (1981). Dissolved gases: some applications to groundwater research. *Transactions of the Geological Society of South Africa* **84**, 91–7.

Heaton, T. H. E., Talma, A. S., and Vogel, J. C. (1983). Origin and history of nitrate in confined groundwater in the western Kalahari. *Journal of Hydrology* **62**, 243–62.

Heaton, T. H. E., Talma, A. S., and Vogel, J. C. (1986*a*). Dissolved gas paleotemperatures and oxygen-18 variations derived from groundwater near Uitenhage, South Africa. *Quaternary Research* **25**, 79–88.

Heaton, T. H. E., Vogel, J. C., Von la Chevallerie, G., and Collett, G. (1986*b*). Climatic influence on the isotopic composition of bone nitrogen. *Nature* **322**, 822–3.

Hecht, A. D., Barry, R., Fritts, M., Imbrie, J., Kutzbach, J., Mitchell, J. M., and Savin, S. M. (1979). Palaeoclimatic research: status and opportunities. *Quaternary Research* **12**, 6–17.

Heine, K. (1978). Radiocarbon chronology of Late Quaternary lakes in the Kalahari, southern Africa. *Catena* **5**, 145–9.

Heine, K. (1981). Aride und pluviale Bedingungen wahrend der letzen Kaltzeit in der Südwes-Kalahari (Südliches Afrika). *Zeitschrift für Geomorphology* (Suppl.) **38**, 1–37.

Heine, K. (1982). The main stages of the Late Quaternary evolution of the Kalahari region, southern Africa. *Palaeoecology of Africa* **15**, 53–76.

Heine, K. (1984). Jungquatare Klimaschwankungen auf der Süddhalbkugel. *Zentralblatt für Geologie und Paläontologie* **1**, 1751–68.

Heine, K. (1985). Late Quaternary development of the Kuiseb River valley and adjacent areas, central Namib Desert, South West Africa/Namibia, and palaeoclimatic implications. *Zeitschrift für Gletscherkunde und Glazialgeologie* **21**, 151–7.

Heine, K. and Geyh, M. A. (1984). Radiocarbon dating of speleothems from the Rössing cave, Namib Desert, and palaeoclimatic implications. In *Late Cainozoic Palaeoclimates of the Southern Hemisphere* (ed. J. C. Vogel), pp. 465–70. Balkema, Rotterdam.

Helgren, D. M. (1977). Geological context of the Vaal River faunas. *South African Journal of Science* **73**, 303–7.

Helgren, D. M. (1978*a*). Environmental history in the north-west Kalahari—a preliminary report. *Palaeoecology of Africa* **10**, 65–6.

Helgren, D. M. (1978*b*). Environmental stratigraphy of the relict alluvia and terraces along the lower Vaal River, South Africa. *Palaeoecology of Africa* **10**, 163–70.

Helgren, D. M. (1979). *Rivers of diamonds: an alluvial history of the lower Vaal basin, South Africa*, University of Chicago Department of Geography Research

Paper 185. University of Chicago.

Helgren, D. M. (1984). Historical geomorphology and geoarchaeology in the southwestern Makgadikgadi basin, Botswana. *Annals of the Association of American Geographers* **74**, 298–307.

Helgren, D. M. and Butzer, K. W. (1977). Paleosols of the southern Cape coast, South Africa: implications for laterite definition, genesis and age. *Geographical Review* **67**, 430–45.

Helgren, D. M. and Brooks, A. S. (1983). Geoarchaeology at /Gi, a Middle Stone Age and Later Stone Age site in the northwestern Kalahari. *Journal of Archaeological Science* **10**, 181–97.

Hendey, Q. B. (1974). Faunal dating of the Late Cenozoic of southern Africa, with special reference to the Carnivora. *Quaternary Research* **4**, 149–61.

Hendey, Q. B. and Cooke, H. B. S. (1985). *Kolpochoerus paiceae* (Mammalia, Suidae) from Skurwerug, near Saldanha, South Africa, and its palaeoenvironmental implications. *Annals of the South African Museum* **97**, 9–56.

Hendey, Q. B. and Singer, R. (1965). The faunal assemblages from the Gamtoos Valley shelters. *South African Archaeological Bulletin* **20**, 206–13.

Hendey, Q. B. and Volman, T. P. (1986). Last Interglacial sea levels and coastal caves in the Cape Province, South Africa. *Quaternary Research* **25**, 189–98.

Hennig, G. J., Grun, R., and Brunnacker, K. (1983). Speleothems, travertines and paleoclimates. *Quaternary Research* **20**, 1–29.

Hobday, D. K. (1976). Quaternary sedimentation and development of the lagoonal complex, Lake St Lucia, Zululand. *Annals of the South African Museum* **71**, 93–113.

Hobday, D. K. and Jackson, M. P. A. (1979). Transgressive shorezone sedimentation and syndepositional deformation in the Pleistocene of Zululand, South Africa. *Journal of Sedimentary Petrology* **49**, 145–8.

Hobday, D. K. and Orme, A. R. (1974). The Port Durnford formation: a major Pleistocene barrier—lagoonal complex along the Zululand coast. *Transactions of the Geological Society of South Africa* **77**, 141–9.

Horowitz, A., Sampson, C. G., Scott, L. and Vogel, J. C. (1978). Analysis of the Voigtspost site, O.F.S., South Africa. *South African Archaeological Bulletin* **33**, 152–9.

Howard, W. R. (1985). Late Quaternary Southern Indian Ocean circulation. *South African Journal of Science* **81**, 253–4.

Howard, W. R. and Prell, W. L. (1984). A comparison of radiolarian and foraminiferal paleoecology in the southern Indian Ocean: new evidence for the interhemispheric timing of climatic change. *Quaternary Research* **21**, 244–63.

Howell, N. (1979). *Demography of the Dobe !Kung.* Academic Press, New York.

Illenberger, W. K. (1986). The Alexandria coastal dunefield: morphology, sand budget and history. Unpublished MSc. thesis, University of Port Elizabeth, Port Elizabeth.

Imbrie, J. and Imbrie, K. P. (1979). *Ice Ages: Solving the Mystery.* London, Macmillan.

Jackson, S. P. (1961). *Climatological Atlas of Africa.* CCTA/CSA, Nairobi.

Jackson, S. P. and Tyson, P. D. (1971). Aspects of weather and climate over southern Africa. *Occasional Paper of the Department of Geography and Environmental Studies.* **6**, 1–13. University of Witwatersrand, Johannesburg.

Jennings, J. N. (1968). A revised map of the desert dunes of Australia. *Australian Geographer* **10**, 408–9.

Kaufman, A. (1986). The distribution of $^{230}Th/^{234}U$ ages in corals and the number of Last Interglacial high-sea stands. *Quaternary Research* **25**, 55–62.

Kent, L. E. (1947). Diatomaceous deposits in the Union of South Africa, with special reference to Kieselguhr. Part 1. *Geological Survey of South Africa Memoirs*, **42**, 1–184.

Kent, L. E. and Gribnitz, K. H. (1985). Freshwater shell deposits in the northwestern Cape Province: further evidence for a widespread wet phase during the Late Pleistocene in South Africa. *South African Journal of Science* **81**, 361–70.

Klammer, G. (1982). Die palaeowuste des Pantanal von Mato Grosso und die Pleistozane Klimageschichte der brasilianischen Randtropen. *Zeitschrift für Geomorphologie NF* **26**, 393–416.

Klein, R. G. (1972*a*). Preliminary report on the July through September 1970 excavations at Nelson Bay Cave, Plettenberg Bay (Cape Province, South Africa). *Palaeoecology of Africa* **6**, 177–208.

Klein, R. G. (1972*b*). The Late Quaternary mammalian fauna of Nelson's Bay Cave (Cape Province, South Africa): its implications for megafaunal extinctions and environmental and cultural change. *Quaternary Research* **2**, 135–42.

Klein, R. G. (1975). Palaeoanthropological implications of the non-archaeological bone assemblage from Swartklip 1, south-western Cape Province, South Africa. *Quaternary Research* **5**, 275–88.

Klein, R. G. (1976). The mammalian fauna of the Klasies River Mouth sites, southern Cape Province, South Africa. *South African Archaeological Bulletin* **31**, 74–98.

Klein, R. G. (1977). The mammalian fauna from the Middle and Later Stone Age (Upper Pleistocene) levels of Border Cave, Natal Province, South Africa. *South African Archaeological Bulletin* **32**, 14–27.

Klein, R. G. (1978*a*). The vertebrate fauna from the Buffelskloof rock shelter. *South African Archaeological Bulletin* **33**, 35–8.

Klein, R. G. (1978*b*). A preliminary report on the larger mammals from the Boomplaas Stone Age cave site, Cango Valley, Oudtshoorn district, South Africa. *South African Archaeological Bulletin* **33**, 66–75.

Klein, R. G. (1980). Environmental and ecological implications of large mammals from Upper Pleistocene and Holocene sites in southern Africa. *Annals of the South African Museum* **81**, 223–83.

Klein, R. G. (1983). Palaeoenvironmental implications of Quaternary large mammals in the fynbos region. In *Fynbos palaeoecology: a preliminary synthesis.* South African National Scientific Progress Report 75 (ed. H. J. Deacon, Q. B. Hendey, and J. J. N. Lambrechts), pp. 116–38. CSIR, Pretoria.

Klein, R. G. (1984). The large mammals of southern Africa: Late Pliocene to Recent. In *Southern African Prehistory and Paleoenvironments* (ed. R. G. Klein), pp. 107–46. Balkema, Rotterdam.

Klein, R. G. (1986). Carnivore size and Quaternary climatic change in southern Africa. *Quaternary Research* **26**, 153–70.

Korn, H. and Martin, H. (1957). The Pleistocene in South West Africa. In *Proceedings of the Third Pan-African Congress on Prehistory, Livingstone 1955* (ed. J. D. Clark), pp. 14–22. Chatto & Windus, London.

Krige, L. J. (1932). The problem of past climates. *South African Journal of Science* **29**, 205–71.

Krishnamurthy, R. V. and Epstein, S. (1985). Tree ring D/H ratio from Kenya, East Africa and its palaeoclimatic significance. *Nature* **317**, 160–2.

Kukla, G., Berger, A., Lotti, R., and Brown, J. (1981). Orbital signature of interglacials. *Nature* **290**, 295–300.

Kuman, K. (1985). The Middle Stone Age at Gi: site formation site disturbance and context. Unpublished paper presented at the Southern African Association of Archaeologists Biennial meeting, Grahamstown, September 1985.

Kuman, K. and Clarke, R. J. (1986). Florisbad—new investigations at a Middle Stone Age hominid site in South Africa. *Geoarchaeology* **1**, 103–125.

Labeyrie, L. D., Pichon, J. J., Labracherie, M., Ippolito, P., Duprat, J., and Duplessy, J. C. (1986). Melting history of Antarctica during the past 60,000 years. *Nature* **322**, 701–6.

Lamb, H. H. (1972). *Climate: Present, Past and Future. 1: Fundamentals and Climate Now.* Methuen, London.

Lambrechts, J. J. N. (1983). Soils, soil processes and soil distribution in the fynbos region: an introduction. In *Fynbos palaeoecology: a preliminary synthesis.* South African National Scientific Progress Report 75 (ed. H. J. Deacon, Q. B. Hendey and J. J. N. Lambrechts), pp. 61–9. CSIR, Pretoria.

Lancaster, I. N. (1978*a*). The pans of the southern Kalahari, Botswana. *Geographical Journal* **144**, 81–98.

Lancaster, I. N. (1978*b*). Composition and formation of southern Kalahari pan margin dunes. *Zeitschrift für Geomorphologie* **22**, 148–69.

Lancaster, I. N. (1979). Evidence for a widespread late Pleistocene humid period in the Kalahari. *Nature* **279**, 145–6.

Lancaster, N. (1981). Palaeoenvironmental implications of fixed dune systems in southern Africa. *Palaeogeography, Palaeoclimatology, and Palaeoecology* **33**, 327–46.

Lancaster, N. (1984*a*). Palaeoenvironments in the Tsondab Valley, central Namib Desert. *Palaeoecology of Africa* **16**, 411–19.

Lancaster, N. (1984*b*). Aridity in southern Africa: age, origins and expression in landforms and sediments. In *Late Cainozoic Palaeoclimates of the Southern Hemisphere* (ed. J. C. Vogel), pp. 433–44. Balkema, Rotterdam.

Lancaster, N. (1986). Pans in the southwestern Kalahari: a preliminary report. *Palaeoecology of Africa* **17**, 59–68.

Lancaster, N. (1987). Formation and reactivation of dunes in the southwestern Kalahari: palaeoclimatic implications. *Palaeoecology of Africa* **18**, in press.

Lancaster, J., Lancaster, N., and Seely, M. K. (1984). The climate of the central Namib Desert. *Madoqua* **14**, 5–61.

Leakey, L. S. B. (1931). *The Stone Age Cultures of Kenya Colony.* Cambridge

University Press, Cambridge.

Leakey, L. S. B. and Solomon, J. D. (1929). East African archaeology. *Nature* **124**, 9.

Leavitt, S. W. and Long, A. (1984). Sampling strategy for stable carbon isotope analysis of tree rings in pine. Nature **311**, 145–7.

Lee, R. B. (1963). The population ecology of man in the early Upper Pleistocene of southern Africa. *Proceedings of the Prehistoric Society* **29**, 235–57.

Lewis, A. D. (1936). Sand dunes of the Kalahari within the Union. *South African Geographical Journal* **19**, 23–32.

Linton, D. L. (1969). Evidence of Pleistocene cryonival phenomena in South Africa. *Palaeoecology of Africa* **5**, 71–89.

Lorius, C., Merlivat, L., Jouzel, J., and Pourchet, M. (1979). A 30,000-yr isotope climatic record from Antarctic ice. *Nature* **280**, 644–8.

Lorius, C., Jouzel, J., Ritz, C., Merlivat, L., Barkov, N. I., Korotkevich, Y. S., and Kotlyakov, V. M. (1985). A 150,000-year climatic record from Antarctic ice. *Nature* **316**, 591–6.

Mabbutt, J. A. (1952). The evolution of the middle Ugab valley, Damaraland, S. W. A. *Transactions of the Royal Society of South Africa* **33**, 334–66.

Mabbutt, J. A. (1957). Some Quaternary events in the winter rainfall area of the Cape Province. In *Proceedings of the Third Pan-African Congress on Prehistory, Livingstone 1955* (ed. J. D. Clark), pp. 6–13. Chatto and Windus, London.

Maggs, T. (1982). Mabhija: pre-colonial industrial development in the Tugela Basin. *Annals of the Natal Museum* **25**, 123–41.

Mallick, D. I. J., Habgood, F., and Skinner, A. C. (1981). A geological interpretation of LANDSAT imagery and air photography of Botswana. *Overseas Geology & Mineralogical Research* **56**, 1–35.

Manabe, S. and Hahn, D. G. (1977). Simulation of the tropical climate of an Ice Age. *Journal of Geophysical Research* **82**, 3889–911.

Marker, M. E. (1972). Karst landform analysis as evidence for climatic change in the Transvaal. *South African Geographical Journal* **54**, 152–62.

Marker, M. E. (1977*a*). A long return geomorphic event in the Namib Desert, South West Africa. *Area* **9**, 209–13.

Marker, M. E. (1977*b*). Aspects of the geomorphology of the Kuiseb River, S. W. A. *Madoqua* **10**, 199–206.

Marker, M. E. and Muller, D. (1978). Relict vlei silts of the middle Kuiseb River, South West Africa. *Madoqua* **11**, 151–62.

Marker, M. E. and Whittington, G. (1971). Observations on some valley forms in the Sani Pass area of Lesotho. *South African Geographical Journal* **53**, 447–65.

Martin, A. K. (1981). Evolution of the Agulhas Current and its palaeo-ecological implications. *South African Journal of Science* **77**, 547–54.

Martin, A. R. H. (1962). Evidence relating to the Quaternary history of the Wilderness lakes. *Transactions of the Geological Society of South Africa* **65**, 19–42.

Martin, A. R. H. (1968). Pollen analysis of Groenvlei lake sediments, Knysna

(South Africa). *Reviews in Palaeobotany and Palynology* **7**, 107–44.

Mason, R. J. (1962). *Prehisotry of the Transvaal.* Witwatersrand University Press, Johannesburg.

Mason, R. J., Brink, A. B. A., and Knight, K. (1959). Pleistocene climatic significance of calcretes and ferricretes. *Nature* **184**, 568.

Mazor, E. (1972). Palaeotemperatures and other hydrological parameters deduced from noble gasses dissolved in groundwaters, Jordan Rift Valley, Israel. *Geochimica and Cosmochimica Acta* **26**, 1321–36.

Maud, R. R. (1968). Quaternary geomorphology and soil formation in coastal Natal. *Zeitschrift für Geomorphologie* (Suppl.) **7**, 155–99.

Maufe, H. B. (1929). *Outline of the Geology of Southern Rhodesia.* Southern Rhodesian Geological Survey Report 24. Second Edition, Bulawayo.

Maufe, H. B. (1930). Changes of climate in Southern Rhodesia during later geological times. *South African Geographical Journal* **13**, 12–16.

McCarthy, M. J. (1967). Stratigraphic and sedimentological evidence from the Durban region of major sea level movements since the late Tertiary. *Transactions of the Geological Society of South Africa* **70**, 135–65.

Morley, J. J. and Hays, J. D. (1979). Comparison of glacial and interglacial oceanographic conditions in the South Atlantic from variations in calcium carbonate and radiolarian distribution. *Quaternary Research* **12**, 396–408.

Mouton, P. le F. N. (1986). Lizard distribution as palaeoenvironmental indicators in the south-western Cape. *Palaeoecology of Africa* **17**, 231–6.

Netterberg, F. (1969). Ages of calcretes in southern Africa. *South African Archaeological Bulletin* **24**, 88–92.

Netterberg, F. (1978). Dating and correlation of calcretes and other pedocretes. *Transactions of the Geological Society of South Africa* **81**, 379–91.

Newell, R. E., Gould-Stewart, S., and Chung, J. C. (1981). Possible interpretation of palaeoclimatic reconstructions for 18 000 BP in the region 60°N to 60°S, 60°W to 100°E. *Palaeoecology of Africa* **13**, 1–19.

Nicholson, S. E. and Flohn, H. (1980). African environmental and climatic changes and the general atmospheric circulation in Late Pleistocene and Holocene. *Climatic Change* **2**, 313–48.

Nilsson, E. (1932). Quaternary glaciations and pluvial lakes in British East Africa. *Geografiska Annaler* (Stockholm) **13**, 249–349.

Nilsson, E. (1940). Ancient changes of climate in British East Africa and Abyssinia. *Geografriska Annaler* (Stockholm) **22**, 1–79.

Ollier, C. D. (1977). Outline geological and geomorphic history of the central Namib Desert. *Madoqua* **10**, 207–12.

Opperman, H. (1978). Excavations in the Buffelskloof rock shelter near Calitzdorp, southern Cape. *South African Archaeological Bulletin* **33**, 18–28.

Opperman, H. (1982). Some research results of excavations in the Colwinton rock shelter, north-eastern Cape. *South African Archaeological Bulletin* **37**, 51–6.

Opperman, H. (1984). 'n Argeologiese ondersoek van noordoos-Kaapland: die bestaansisteem van jagter-versamelaars uit die Later Steentydperk. Unpublished DPhil thesis, University of Stellenbosch.

Orme, A. R. (1976). Late Pleistocene channels and Flandrian sediments beneath Natal estuaries. *Annals of the South African Museum* **71**, 77–85.

Parkington, J. (1981). The effects of environmental change on the scheduling of visits to the Elands Bay Cave, Cape Province, South Africa. In *Pattern of the Past: Studies in Honour of David Clarke* (ed. I. Hodder, G. Isaac, and N. Hammond), pp. 341–59. Cambridge University Press, Cambridge.

Parkington, J. (1986). Landscape and subsistence changes since the Last Glacial Maximum along the western Cape coast. In *The End of the Paleolithic in the Old World* (ed. L. G. Straus), pp. 201–27. BAR International Series 284, Oxford.

Partridge, T. C. (1978). Reappraisal of the stratigraphy of the Sterkfontein hominid site. *Nature* **275**, 282–7.

Partridge, T. C. and Brink, A. B. A. (1967). Gravels and terraces of the lower Vaal River basin. *South African Geographical Journal* **49**, 21–38.

Partridge, T. C. and Dalbey, T. S. (1986). Geoarchaeology of the Haaskraal Pan: a preliminary palaeoenvironmental model. *Palaeoecology of Africa* **17**, 69–78.

Peabody, F. E. (1954). Travertines and cave deposits of the Kaap escarpment of South Africa and the type locality of *Australopithecus africanus* Dart. *Bulletin of the Geological Society of America* **65**, 671–706.

Penck, A. (1905). Climatic features of the Pleistocene Ice Age. *South African Journal of Science* **2**, 1–9.

Penck, A. (1914). The shifting of the climatic belts. *Scottish Geographical Magazine* **1914**, 281–3.

Pether, J. (1986). Late Tertiary and early Quaternary marine deposits of the Namaqualand coast, Cape Province: new perspectives. *South African Journal of Science* **82**, 464–70.

Petrov, M. P. (1975). *Deserts of the World.* Wiley, New York.

Pierce, S. M. (1984). A synthesis of plant phenology in the Fynbos Biome. *South African National Scientific Programmes Report* **88**, 1–55. CSIR, Pretoria.

Pierce, S. M. and Cowling, R. M. (1984). Phenology of fynbos, renosterveld and subtropical thicket in the south-eastern Cape. *South African Journal of Botany* **3**, 1–16.

Pittock, A. B. (1973). Global meridional interactions in the stratosphere and troposphere. *Quarternary Journal of the Royal Meteorological Society* **99**, 424–37.

Pittock, A. B. (1978). An overview. In *Climatic Change and Variability: a Southern Perspective* (ed. A. B. Pittock, L. A. Frakes, D. Jenssen, J. A. Peterson, and J. W. Zillman), pp. 1–8. Cambridge University Press, Cambridge.

Pittock, A. B. and Salinger, M. J. (1982). Towards regional scenarios for a CO_2-warmed earth. *Climatic Change* **4**, 23–40.

Poynton, R. J. (1971). A silvicultural map of southern Africa. *South African Journal of Science* **67**, 58–60.

Poynton, R. J. (1986). Historical biogeography: theme and South African variations. *Palaeoecology of Africa* **17**, 139–53.

Prell, W. L., Hutson, W. H., and Williams, D. F. (1979). The Subtropical Convergence and late Quaternary circulation in the southern Indian Ocean. *Marine*

Micropalaeontology **4**, 225–34.

Price-Williams, D., Watson, A., and Goudie, A. S. (1982). Quaternary colluvial stratigraphy, archaeological sequences and palaeoenvironments in Swaziland, southern Africa. *Geographical Journal* **148**, 50–67.

Prior, J. (1984). Investigations on archaeological charcoals from Swaziland, using SEM techniques. In *Late Cainozoic Palaeoclimates of the Southern Hemisphere* (ed. J. C. Vogel), pp. 353–60. Balkema, Rotterdam.

Prior, J. and Price-Williams, D. (1985). An investigation of climatic change in the Holocene Epoch using archaeological charcoal from Swaziland, southern Africa. *Journal of Archaeological Science* **12**, 457–75.

Rogers, A. W. (1905). A raised beach deposit near Klein Brak River. *Annual Report of the Geological Commission of the Cape of Good Hope* **10**, 293–6.

Rogers, A. W. (1922). Post-Cretaceous climates of South Africa. *South African Journal of Science* **19**, 2–31.

Rogers, J. (1982). Lithostratigraphy of Cenozoic sediments between Cape Town and Elands Bay. *Palaeoecology of Africa* **15**, 121–37.

Rogers, J. (1986). Cenozoic geology of the southern Cape coastal plain, particularly the area between the Kafferkuils and Gouritz estuaries. *Palaeoecology of Africa* **17**, 3–12.

Rognon, P. (1982). Pluvial and arid phases in the Sahara: the role of non-climatic factors. *Palaeoecology of Africa* **12**, 45–62.

Ruddiman, W. F. and Duplessy, J. C. (1985). Conference on the Last Deglaciation: timing and mechanism. *Quaternary Research* **23**, 1–17.

Ruddock, A. (1968). Cainozoic sea levels and diastrophism in a region bordering on Algoa Bay. *Transactions of the Geological Society of South Africa* **71**, 209–33.

Rust, U. (1984). Geomorphic evidence of Quaternary climatic changes in Etosha, South West Africa/Namibia. In *Late Cainozoic Palaeoclimates of the Southern Hemisphere* (ed. J. C. Vogel), pp. 279–86. Balkema, Rotterdam.

Rust, U. and Wienecke, F. (1974). Studies on gramadulla formation in the middle part of the Kuiseb River, South West Africa. *Madoqua* **3**, 5–15.

Rust, U. and Wienecke, F. (1980). A reinvestigation of some aspects of the Kuiseb valley upstream of Gobabeb, South West Africa. *Madoqua* **12**, 163–73.

SACS (South African Committee for Stratigraphy) (1980). *Stratigraphy of South Africa. Part I: Lithostratigraphy of the Republic of South Africa, SWA/Namibia and the Republics of Bophuthatswana, Transkei and Venda.* Geological Society of South Africa Handbook 8, Pretoria.

Salinger, M. J. (1981). Palaeoclimates north and south. *Nature* **291**, 106–7.

Salinger, M. J. (1984). New Zealand climate: the last 5 million years. In *Late Cainozoic Palaeoclimates of the Southern Hemisphere* (ed. J. C. Vogel), pp. 131–50. Balkema, Rotterdam.

Sarnthein, M. (1978). Sand deserts during glacial maximum and climatic optimum. *Nature* **272**, 43–5.

Schalke, H. J. W. G. (1973). The upper Quaternary of the Cape Flats area (Cape Province, South Africa). *Scripta Geologica* **15**, 1–57.

Scholtz, A. (1986). Palynological and palaeobotanical studies in the southern

Cape. Unpublished MA thesis, University of Stellenbosch.

Schulze, B. R. (1972). South Africa. In *Climates of Africa. World Survey of Climatology*, Vol. 10 (ed. J. F. Griffiths), pp. 501–66. Elsevier, Amsterdam.

Schulze, B. R. (1984). *Climate of South Africa, Part 8. General Survey WB28*, 5th edn. South African Weather Bureau, Pretoria.

Schulze, R. E. and McGee, O. S. (1978). Climatic indices and classification in relation to the biogeography of southern Africa. In *Biogeography and Ecology of Southern Africa* (ed. M. J. A. Werger), pp. 19–52. W. Junk, The Hague.

Schumm, S. A. (1968). River adjustment to altered hydrologic regime—Murrumbigee River and palaeochannels, Australia. *US Geological Survey Professional Paper* **598**, 1–65.

Schweitzer, F. and Wilson, M. L. (1982). Byneskranskop 1. A late Quaternary living site in the southern Cape Province, South Africa. *Annals of the South African Museum* **88**, 1–203.

Scott, L. (1982*a*). A Late Quaternary pollen record from the Transvaal Bushveld, South Africa. *Quaternary Research* **17**, 339–70.

Scott, L. (1982*b*). Late Quaternary fossil pollen grains from the Transvaal, South Africa. *Reviews in Palaeobotany and Palynology* **36**, 241–78.

Scott, L. (1984*a*). Palynological evidence for Quaternary palaeoenvironments in southern Africa. In *Southern African Palaeoenvironments and Prehistory* (ed. R. G. Klein), pp. 65–80. Balkema, Rotterdam.

Scott, L. (1984*b*). Late Quaternary palaeoenvironments in the Transvaal on the basis of palynological evidence. In *Late Cainozoic Palaeoclimates of the Southern Hemisphere* (ed. J. C. Vogel), pp. 317–28. Balkema, Rotterdam.

Scott, L. (1986). Pollen analysis and palaeoenvironmental interpretation of Late Quaternary sediment exposures in the eastern Orange Free State, South Africa. *Palaeoecology of Africa* **17**, 113–22.

Scott, L. and Klein, R. G. (1981). A hyena-accumulated bone assemblage from late Holocene deposits at Deelpan, Orange Free State. *Annals of the South African Museum* **86**, 217–27.

Scott, L. and Vogel, J. C. (1983). A late Quaternary pollen profile from the Transvaal Highveld, South Africa. *South African Journal of Science* **79**, 266–72.

Sealy, J. and Van der Merwe, N. J. (1986). Isotope assessment and the seasonal mobility hypothesis in the southwestern Cape of South Africa. *Current Anthropology* **27**, 135–49.

Seely, M. K. (1978). The Namib Dune Desert: an unusual ecosystem. *Journal of Arid Environments* **1**, 117–28.

Seely, M. K. and Louw, G. N. (1980). First approximation of the effects of rainfall on the ecology and energetics of a Namib Desert dune ecosystem. *Journal of Arid Environments* **3**, 25–54.

Seely, M. K. and Sandelowsky, B. H. (1974). Dating the regression of a river's end point. *South African Archaeological Society Goodwin Series* **2**, 61–4.

Selby, M. J., Hendy, C. H. and Seely, M. K. (1979). A late Quaternary lake in the central Namib Desert, southern Africa, and some implications. *Palaeogeography, Palaeoclimatology, and Palaeoecology* **26**, 37–41.

Shackleton, N. J. (1973). Oxygen isotope analysis as a means of determining season of occupation of prehistoric midden sites. *Archaeometry* **15**, 133–41.

Shackleton, N. J. (1982). Stratigraphy and chronology of the Klasies River Mouth deposits: oxygen isotope evidence. In *The Middle Stone Age at Klasies River Mouth in South Africa* (ed. R. Singer and J. Wymer), pp. 194–9. Chicago University Press, Chicago.

Shackleton, N. J. and Opdyke, N. D. (1976). Oxygen-isotope paleomagnetic stratigraphy of Pacific Core V28-239 late Pliocene to latest Pleistocene. *Geological Society of America Memoirs* **145**, 449–64.

Shackley, M. (1980). An Acheulian industry with *Elephas recki* fauna from Namib IV, South West Africa (Namibia). *Nature* **284**, 340–1.

Shackley, M. (1985). Palaeolithic archaeology of the central Namib Desert. *Cimbebasia (B) Memoir* **6**, 1–84.

Shaw, P. (1986*a*). The palaeohydrology of the Okavango delta—some preliminary results. *Palaeoecology of Africa* **17**, 51–8.

Shaw, P. (1986*b*). Geomorphic evidence for the Late Quaternary palaeoclimates of the middle Kalahari of northern Botswana. *Catena* **13**, 349–59.

Siesser, W. (1980). Late Miocene origin of the Benguela upwelling system off northern Namibia. *Science* **208**, 283–5.

Singer, R. and Wymer, J. (1982). *The Middle Stone Age at Klasies River Mouth in South Africa.* Chicago University Press, Chicago.

Smuts, J. C. (1932). Climate and man in Africa. *South African Journal of Science* **29**, 98–131.

Sohnge, P. G., Visser, D. J. L., and Van Riet Lowe, C. (1937). *The Geology and Archaeology of the Vaal River Basin*, Geological Survey Memoirs, No. 35. Government Printer, Pretoria.

Solomon, J. D. (1939). The Pleistocene succession in Uganda. In *The Prehistory of the Uganda Protectorate* (ed. T. P. O'Brien), pp. 15–50. Cambridge University Press, Cambridge.

Sparrow, G. W. A. (1967). Pleistocene periglacial topography in southern Africa. *Journal of Glaciology* **6**, 551–9.

Sparrow, G. W. A. (1971). Some Pleistocene studies in southern Africa. *Journal of Geography* **3**, 809–14.

Sprigg, R. C. (1979). Stranded and submerged sea-beach systems of southeast South Australia and the aeolian desert cycle. *Sedimentary Geology* **22**, 53–96.

Street, F. A. and Grove, A. T. (1976). Environmental and climatic implications of late Quaternary lake level fluctuations in Africa. *Nature* **261**, 385–90.

Street, F. A. and Grove, A. T. (1979). Global maps of lake level fluctuations since 30,000 years B.P. *Quaternary Research* **12**, 83–118.

Street-Perrott, F. A. and Harrison, S. P. (1985). Lake levels and climate reconstruction. In *Paleoclimate Analysis and Modeling* (ed. A. D. Hecht), pp. 291–340. Wiley, New York.

Stuckenberg, B. R. (1969). Effective temperature as an ecological factor in southern Africa. *Zoologica Africana* **4**, 145–97.

Summerfield, M. A. (1983). Silcrete as a palaeoclimatic indicator. *Palaeogeo-*

graphy, Palaeoclimatology, and Palaeoecology **41**, 65–80.

Summers, R. F. (1960). Environment and culture in Southern Rhodesia. *Proceedings of the American Philosophical Society* **104**, 266–91.

Szabo, B. J. and Butzer, K. W. (1979). Uranium series dating of lacustrine limestones from pan deposits with final Acheulian assemblage at Rooidam, Kimberley district, South Africa. *Quaternary Research* **11**, 257–60.

Talbot, M. R. (1980). Environmental responses to climatic change in the West African Sahel over the past 20,000 years. In *The Sahara and the Nile* (ed. M. A. J. Williams and H. Faure), pp. 37–62. Balkema, Rotterdam.

Talbot, M. R. (1984). Late Pleistocene dune building and rainfall in the Sahel. *Palaeoecology of Africa* **16**, 203–14.

Taljaard, J. J. (1981). Upper air circulation, temperature and humidity over southern Africa. *South African Weather Bureau Technical Paper* **9**, 1–92.

Talma, A. S., Vogel, J. C., and Partridge, T. C. (1974). Isotopic contents of some Transvaal speleothems and their palaeoclimatic significance. *South African Journal of Science* **70**, 135–40.

Tankard, A. J. (1975). Thermally anomalous Late Pleistocene molluscs from the south-western Cape Province. *Annals of the South African Museum* **69**, 17–45.

Tankard, A. J. (1976*a*). The stratigraphy of a coastal cave and its palaeoclimatic significance. *Palaeoecology of Africa* **9**, 151–9.

Tankard, A. J. (1976*b*). Pleistocene history and coastal morphology of the Yserfontein-Elands Bay area, Cape Province. *Annals of the South African Museum* **69**, 73–119.

Tankard, A. J. and Rogers, J. (1978). Late Cenozoic palaeoenvironments on the west coast of southern Africa. *Journal of Biogeography* **5**, 319–37.

Tankard, A. J. and Schweitzer, F. R. (1974). The geology of Die Kelders Cave and environs—a palaeoenvironmental study. *South African Journal of Science* **70**, 356–69.

Tankard, A. J., Jackson, M. P. A., Eriksson, K. A., Hobday, D. K., Hunter, D. R., and Minter, W. E. L. (1982). *Crustal Evolution of Southern Africa: 3.8 Billion Years of Earth History.* Springer Verlag, New York.

Teller, J. T. and Lancaster, N. (1985). History of sediments at Khommabes, central Namib desert. *Madoqua* **14**, 267–78.

Teller, J. T. and Lancaster, N. (1986*a*). Interdune lacustrine deposits of the Namib Sand Sea, Namibia. Abstract, 12th International Congress on Sedimentology, Canberra.

Teller, J. T. and Lancaster, N. (1986*b*). Lacustrine sediments at Narabeb in the central Namib Desert, Namibia. *Palaeogeography, Palaeoclimatology, and Palaeoecology* **56**, 177–95.

Thackeray, J. F. (1984). Climatic change and mammalian fauna from Holocene deposits in Wonderwerk Cave, northern Cape. In *Late Cainozoic Palaeoclimates of the Southern Hemisphere* (ed. J. C. Vogel), pp. 371–4. Balkema, Rotterdam.

Thackeray, J. F. (1987). Late Quaternary environmental changes inferred from small mammalian fauna, southern Africa. *Climatic Change* **10**, 285–305.

Thomas, D. S. G. (1984). Ancient ergs of the former arid zones of Zimbabwe,

Zambia and Angola. *Transactions of the Institute of British Geographers N.S.* **9**, 75–88.

Thomas, D. S. G. and Goudie, A. S. (1984). Ancient ergs of the southern hemisphere. In *Late Cainozoic Palaeoclimates of the Southern Hemisphere* (ed. J. C. Vogel), pp. 407–18. Balkema, Rotterdam.

Tomlinson, R. W. (1974). Preliminary biogeographical studies on the Inyanga mountains, Rhodesia. *South African Geographical Journal* **56**, 15–26.

Turon, J. L. (1984). Direct land-sea correlations in the Last Interglacial complex. *Nature* **309**, 673–6.

Tusenius, M. L. (1986). The study of charcoal from some southern African archaeological contexts. Unpublished M.A. thesis, University of Stellenbosch.

Tyson, P. D. (1977). The enigma of changing world climates. *South African Geographical Journal* **59**, 77–116.

Tyson, P. D. (1981). Atmospheric circulation variations and the occurrence of extended wet and dry spells over southern Africa. *Journal of Climatology* **1**, 115–30.

Tyson, P. D. (1984). The atmospheric modulation of extended wet and dry spells over South Africa, 1958–1978. *Journal of Climatology* **4**, 621–35.

Tyson, P. D. (1986). *Climatic Change and Variability in Southern Africa.* Oxford University Press, Cape Town.

Tyson, P. D. and Dyer, T. G. J. (1975). Mean annual fluctuations of precipitation in the summer rainfall region of South Africa. *South African Geographical Journal* **57**, 104–10.

Tyson, P. D. and Dyer, T. G. J. (1978). The predicted above-normal rainfall of the seventies and the likelihood of droughts in the eighties in South Africa. *South African Journal of Science* **74**, 372–7.

Tyson, P. D. and McNaughton, J. (1980). A preliminary assessment of *Podocarpus falcatus* in dendrochronological and dendroclimatological studies in the Witelsbos Forest Reserve. *South African Forestry Journal* **111**, 29–32.

Tyson, P. D., Dyer, T. G. J., and Mametse, M. N. (1975). Secular changes in South African rainfall. *Quarterly Journal of the Royal Meteorological Society* **101**, 817–83.

Van Daalen, J. C. (1980). The colonization of fynbos and disturbed sites by indigenous forest communities in the southern Cape. Unpublished MSc thesis, University of Cape Town.

Van der Merwe, N. J. (1986). Carbon isotope ecology of herbivores and carnivores. *Palaeoecology of Africa* **17**, 123–32.

Van der Merwe, N. J. and Vogel, J. C. (1983). Recent carbon isotope research and its implications for African archaeology. *African Archaeological Review* **1**, 33–56.

Van Riet Lowe, C. (1929). Further notes on the archaeology of Sheppard Island. *South African Journal of Science* **26**, 623–30.

Van Riet Lowe, C. (1930). Fresh light on the prehistoric archaeology of South Africa. *Bantu Studies* **3–4**, 385–94.

Van Riet Lowe, C. (1937). The archaeology of the Vaal River basin. *Geological Survey Memoirs* **35**, 61–164.

Van Rooyen, T. H. and Burger, R. du T. (1974). Dispersal and sedimentary characteristics of sandy soil parent materials in semi-arid South Africa. *Zeitschrift für Geomorphologie* Suppl. **20**, 69–90.

Van Zinderen Bakker, E. M. (1951). Archaeology and palynology. *South African Archaeological Bulletin* **6**, 1–8.

Van Zinderen Bakker, E. M. (1955). A preliminary survey of the peat bogs of the Alpine belt of northern Basutoland. *Acta Geographica* **14**, 413–22.

Van Zinderen Bakker, E. M. (1957). Pollen analytical investigation of the Florisbad deposits. In *Proceedings of the Third Pan-African Congress on Prehistory, Livingstone 1955* (ed. J. D. Clark), pp. 56–67. Chatto and Windus, London.

Van Zinderen Bakker, E. M. (1962*a*). Pleistocene climates and cultures in northeastern Angola. *Nature* **196**, 639–42.

Van Zinderen Bakker, E. M. (1962*b*). Botanical evidence for Quaternary climates in Africa. *Annals of the Cape Provincial Museums* **2**, 16–31.

Van Zinderen Bakker, E. M. (1967). Upper Pleistocene and Holocene stratigraphy and ecology on the basis of vegetation changes in sub-Saharan Africa. In *Background to Evolution in Africa* (ed. W. W. Bishop and J. D. Clark), pp. 125–47. University of Chicago Press, Chicago.

Van Zinderen Bakker, E. M. (1975). Origin and palaeoenvironments of the Namib Desert biome. *Journal of Biogeography* **2**, 65–73.

Van Zinderen Bakker, E. M. (1976). The evolution of Late Quaternary palaeoclimates of southern Africa. *Palaeoecology of Africa* **9**, 160–202.

Van Zinderen Bakker, E. M. (ed.) (1978). *Antarctic Glacial History and World Palaeoenvironments.* Balkema, Rotterdam.

Van Zinderen Bakker, E. M. (1980). Comparison of late Quaternary climatic evolutions in the Sahara and the Namib–Kalahari region. *Palaeoecology of Africa* **12**, 381–94.

Van Zinderen Bakker, E. M. (1982*a*). African palaeoenvironments 18 000 yrs B.P. *Palaeoecology of Africa* **15**, 77–99.

Van Zinderen Bakker, E. M. (1982*b*). Pollen analytical studies of the Wonderwerk Cave, South Africa. *Pollen et Spores* **24**, 235–50.

Van Zinderen Bakker, E. M. (1984). Aridity along the Namibian coast. *Palaeoecology of Africa* **16**, 149–60.

Van Zinderen Bakker, E. M. and Butzer, K. W. (1973). Quaternary environmental changes in southern Africa. *Soil Science* **116**, 236–48.

Van Zinderen Bakker, E. M. and Clark, J. D. (1962). Pleistocene climates and culture in northeastern Angola. *Nature* **196**, 639–42.

Van Zinderen Bakker, E. M. and Werger, M. J. A. (1974). Environment, vegetation and phytogeography of the high altitude bogs of Lesotho. *Vegetatio* **29**, 37–49.

Vincent, E. (1972). Climatic change at the Pleistocene-Holocene boundary in the southwestern Indian Ocean. *Palaeoecology of Africa* **6**, 45–54.

Visser, D. J. L. and Van Riet Lowe, C. (1955). The geology and archaeology of the Little Caledon River valley. *Memoirs of the Geological Society of South Africa* **47**, 1–53.

Vogel, J. C. (1970). Groningen radiocarbon dates IX. *Radiocarbon* **12**, 444–71.

Vogel, J. C. (1978). Isotopic assessment of the dietary habits of ungulates. *South African Journal of Science* **74**, 298–301.

Vogel, J. C. (1982). Age of the silt terrace at Homeb. *Palaeoecology of Africa* **15**, 201–9.

Vogel, J. C. (1983). Isotopic evidence for past climates and vegetation of South Africa. *Bothalia* **14**, 391–4.

Vogel, J. C. (ed.) (1984). *Late Cainozoic Palaeoenvironments of the Southern Hemisphere.* Balkema, Rotterdam.

Vogel, J. C. and Marais, M. (1971). Pretoria radiocarbon dates. *Radiocarbon* **13**, 378–94.

Vogel, J. C. and Partridge, T. C. (1984). Preliminary metric ages for the Taung tufas. In *Late Cainozoic Palaeoclimates of the Southern Hemisphere* (ed. J. C. Vogel), pp. 507–14. Balkema, Rotterdam.

Vogel, J. C. and Visser, E. (1981). Pretoria radiocarbon dates II. *Radiocarbon* **23**, 43–80.

Voigt, E. A. (1973). Stone Age molluscan utilization at Klasies River Mouth caves. *South African Journal of Science* **69**, 306–9.

Voigt, E. A. (1982). The molluscan fauna. In *The Middle Stone Age at Klasies River Mouth in South Africa* (ed. R. Singer and J. Wymer), pp. 155–86. University of Chicago Press, Chicago.

Wadley, L. and Verhagen B. Th. (1986). Filling the gap: a new Holocene chronology for the Transvaal Stone Age. *South African Journal of Science* **82**, 271.

Ward, J. D. (1982). Aspects of a suite of Quaternary conglomeratic sediments in the Kuiseb valley, Namibia. *Palaeoecology of Africa* **15**, 211–16.

Ward, J. D. (1984*a*). Aspects of the Cenozoic geology in the Kuiseb Valley, central Namib Desert. Unpublished PhD. thesis, University of Natal.

Ward, J. D. (1984*b*). A reappraisal of the Cenozoic stratigraphy in the Kuiseb valley of the central Namib Desert. In *Late Cainozoic Palaeoclimates of the Southern Hemisphere* (ed. J. C. Vogel), pp. 455–64. Balkema, Rotterdam.

Ward, J. D., Seely, M. K., and Lancaster, N. (1983). On the antiquity of the Namib. *South African Journal of Science* **79**, 175–83.

Warren, A. (1970). Dune trends and their implications in the central Sudan. *Zeitschrift für Geomorphologie* Suppl. **10**, 154–80.

Wasson, R. J. *et al.* (1983). Geomorphology, late Quaternary stratigraphy and palaeoclimatology of the Thar dunefield. *Zeitschrift für Geomorphologie* Suppl. **45**, 117–51.

Wasson, R. J. (1984). Late Quaternary palaeoenvironments in the desert dune-fields of Australia. In *Late Cainozoic Palaeoclimates of the Southern Hemisphere* (ed. J. C. Vogel), pp. 419–32. Balkema, Rotterdam.

Watts, N. L. (1980). Quaternary pedogenic calcretes from the Kalahari (southern Africa), mineralogy, genesis and diagenesis. *Sedimentology* **27**, 661–86.

Wayland, E. J. (1934). Rifts, rivers, rains and early man in Uganda. *Journal of the Royal Anthropological Institute* **64**, 333–52.

Wayland, E. J. (1954). Outline of prehistory and Stone Age climatology in the Bechuanaland Protectorate. *Memoires d'Académie Royale des Sciences Coloniales* **25**, 1–42.

Wellington, J. H. (1945). Notes on the drainage of the western Free State sandveld. *South African Geographical Journal* **28**, 73–7.

Wendt, W. E. (1972). Preliminary report on an archaeological research programme in South West Africa. *Cimbebasia* B **2**(1), 1–61.

Williams, J. (1978). A brief comparison of model simulations of glacial period maximum atmospheric circulation. *Palaeogeography, Palaeoclimatology and Palaeoecology* **25**, 191–8.

Williams, J., Barry, R. G., and Washington, W. M. (1974). Simulation of the atmospheric circulation using the NCAR global circulation model with Ice Age boundary conditions. *Journal of Applied Meteorology* **13**, 305–17.

Wright, E. P. (1978). Geological studies in the northern Kalahari. *Geographical Journal* **144**, 235–49.

Yaalon, D. H. (1982). On the aeolianite-red sands relationship in Natal. *Palaeoecology of Africa* **15**, 145–8.

Yaalon, D. H. and Ward, J. D. (1982). Observations on calcrete and recent calcic horizons in relation to landforms in the central Namib Desert. *Palaeoecology of Africa* **15**, 183–6.

Yates, R. J., Miller, D. E., Halkett, D. J., Manhire, A. H., and Parkington, J. E. (1986). A late mid-Holocene high sea-level: a preliminary report on geoarchaeology at Elands Bay, western Cape Province, South Africa. *South African Journal of Science* **82**, 164–5.

Zeuner, F. (1952). *Dating the Past.* Methuen, London.

Author index

Acocks, J. P. H. 90, 137
Ager, T. A. 34
Aharon, P. 34
Alexandre, J. 33, 96
Allchin, B. 30, 63
Ambrose, S. 43
Armstrong, A. L. 5
Ash, J. E. 27, 30, 64
Avery, D. M. 41, 74, 81, 82, 83, 125, 126, 128, 129, 130, 134, 135, 145, 146, 147, 153, 154, 157, 160, 166, 167, 168, 177, 178, 181, 182

Backman, J. 1
Baillieul, T. A. 60
Barkov, N. I. 157
Barry, R. 2, 148, 161
Barry, R. G. 20
Barwis, J. H. 117
Beaumont, P. B. 9, 10, 33, 46, 53, 58, 73, 74, 75, 76, 79, 80, 81, 82, 102, 154, 155, 156, 159, 164, 165, 166, 167, 175, 178, 181, 182, 184
Berger, A. 154
Bernard, E. A. 7, 8
Birch, G. F. 111, 114, 121
Birkeland, P. W. 35
Bond, G. 31, 63
Boocock, C. 60
Boucher, K. 21
Bowler, J. M. 27, 30, 63, 69
Brain, C. K. 9, 33, 41, 55
Brain, V. 55
Brink, A. B. A. 9, 31, 35, 105
Brink, J. 9, 37, 38, 40, 41, 115, 116, 123, 131, 132, 146, 147, 153, 154, 156, 157, 159, 164, 166, 167, 168, 169, 172, 177
Brook, G. A. 85
Brooks, A. S. 68, 155, 184
Brown, A. J. V. 39
Brown, J. 154
Brunnacker, K. 33
Bružewicz, A. J. 24, 33, 76, 77, 98, 155, 157, 158, 160, 165, 166, 167, 182
Buckley, R. L. 63
Burger, R. du T. 30
Butzer, K. W. 4, 7, 9, 10, 24, 31, 32, 33, 34, 35, 72, 73, 74, 76, 77, 78, 79, 80, 81, 85, 93, 96, 97, 98, 100, 102, 105, 106, 107, 111, 112, 113, 115, 117, 118, 122, 123, 124, 125, 126, 128, 129, 130, 146, 152, 153, 154, 155, 156, 157, 158, 159, 160, 164, 165, 166, 167, 172, 175, 177, 179, 180, 181, 182, 183

Cable, J. H. C. 103
Caratini, C. 37, 50
Carter, P. L. 97, 98, 103, 153, 154, 180
Cartmill, M. 41
Chappell, J. 119, 127, 152, 154, 157, 165
Chung, J. C. 23, 64, 166
Clark, J. D. 5, 45, 66
Clarke, R. J. 100
CLIMAP Project Members 11, 19, 34, 97
Cockcroft, M. J. 24, 27, 156, 174
Coetzee, J. A. 9, 32, 36, 37, 99, 100, 107, 138, 157, 167, 179
Collett, G. 44, 45, 141, 142
Cooke, H. B. S. 5, 6, 7, 39, 115, 116
Cooke, H. J. 9, 32, 33, 66, 67, 69, 154, 160, 164, 165
Corvinus, G. 116
Cowling, R. M. 137, 169, 170
Cronin, T. M. 34
Cruz-Uribe, K. 153
Curtis, B. A. 38

Daitz, L. D. 38, 132, 133, 138, 169, 170, 178
Dalbey, T. S. 32, 57, 155

Dardis, G. F. 96
Dart, R. A. 76
Davies, O. 9, 34, 87, 88, 89, 111
Deacon, H. J. 4, 9, 21, 24, 33, 37, 38, 40, 41, 46, 59, 115, 116, 118, 123, 125, 126, 127, 131, 132, 138, 145, 146, 147, 148, 152, 153, 154, 156, 157, 159, 164, 165, 166, 167, 168, 169, 170, 172, 177, 178
Deacon, J. 9, 37, 38, 40, 41, 46, 64, 71, 93, 123, 124, 131, 132, 146, 147, 153, 154, 156, 157, 159, 164, 166, 167, 168, 169, 172, 177
De Graaff, G. 41
De Niro, M. J. 43
De Villiers, H. 9, 10, 82, 153, 164, 165
Dingle, R. V. 34, 111, 118, 119
Dunwiddie, P. W. 38, 39, 138
Du Plessis, A. 111
Duplessy, J. C. 146, 157, 167
Duprat, J. 146, 157
Du Toit, A. L. 6, 60, 69
Dyer, T. G. J. 15, 24, 38, 96

Ebert, J. I. 67
Einsele, G. 119
Ellenberger, P. 97
Epstein, S. 39
Erikson, J. L. 177
Eriksson, K. A. 60

Fairbridge, R. W. 1, 2
Fairhall, A. W. 177
Fitzpatrick, R. W. 96, 97
Flemming, B. W. 118, 120, 121, 159
Flint, R. F. 4, 6, 7, 29, 63
Flohn, H. 10, 21, 24, 25, 158, 172, 173, 174
Fock, G. J. 31, 32, 105, 107, 157, 179, 180
Fritts, M. 2, 148, 161
Fritz, C. 157
Fryberger, S. G. 30

Gasse, F. 32
Gates, W. L. 19, 20, 27
Geleijnse, V. B. 115, 116, 125, 126, 127, 146, 148, 152, 165
Gevers, T. W. 54
Geyh, M. A. 55, 154, 164, 168, 185
Gibbs Russell, G. E. 168
Gillooly, J. F. 38, 92
Giresse, P. 119
Gooch, W. D. 45
Goodwin, A. J. H. 45
Goudie, A. S. 30, 34, 62, 63, 66, 86, 87, 106
Gould Stewart, S. 23, 64, 166
Granger, J. E. 96
Gregory, J. W. 4
Grey, D. R. C. 32, 66, 69, 160
Gribnitz, K. H. 58, 154, 164, 178
Grindley, J. R. 58
Grine, F. E. 76
Grove, A. T. 7, 30, 32, 62, 63, 65, 66, 67, 172, 184
Grun, R. 33

Habgood, F. 30, 62, 66
Hahn, D. G. 19, 20
Halkett, D. J. 120, 121, 159
Hall, M. 38
Harper, G. 34, 96, 97
Harrison, M. S. J. 10, 16, 167, 174, 175
Harrison, S. P. 32
Hastenrath, S. 33, 96
Hays, J. D. 23, 27, 166
Hazel, J. E. 34
Heath, G. R. 19, 27
Heaton, T. H. E. 36, 44, 45, 64, 72, 86, 97, 141, 142, 143, 153, 155, 159, 160, 164, 167, 178
Hecht, A. D. 2, 148, 161
Hegde, K. T. M. 30, 63
Heine, K. 24, 25, 54, 55, 64, 65, 66, 67, 73, 154, 155, 158, 164, 166, 167, 168, 175, 183, 184, 185
Helgren, D. M. 9, 24, 31, 33, 35, 66, 67, 68, 76, 77, 98, 105, 111, 112, 122, 124, 128, 155, 157, 159, 160, 165, 166, 167, 177, 182, 184
Hendey, Q. B. 9, 39, 88, 115, 116, 117
Hendy, C. H. 50, 53
Hennig, G. J. 33
Herm, D. 119
Hitchcock, R. K. 67
Hobday, D. K. 60, 88, 89

Horowitz, A. 107, 160, 164, 180
Howard, W. R. 27
Howell, N. 172
Hunter, D. R. 60
Hutson, W. H. 166

Illenberger, W. K. 30, 34, 111, 121
Imbrie, J. 2, 9, 148, 161
Imbrie, K. P. 9
Ippolito, P. 146, 157
Isaac, G. L. 7

Jackson, M. P. A. 60, 88
Jackson, S. P. 12, 14, 16
Jennings, J. N. 30, 63
Jones, N. 5
Jouzel, J. 157

Kaufman, A. 34
Kent, L. E. 56, 58, 154, 164, 178
Klammer, G. 63
Klein, R. G. 9, 39, 40, 46, 76, 79, 84, 95, 100, 107, 110, 117, 124, 125, 126, 127, 128, 130, 134, 138, 139, 140, 141, 145, 146, 147, 151, 152, 153, 155, 157, 158, 164, 165, 167, 169, 177
Knight, K. 35
Korn, H. 5, 50, 54
Kouyoumontzakis, G. 119
Korotkevich, Y. S. 157
Kotlyakov, V. M. 157
Krige, L. J. 5
Krishnamurthy, R. V. 39
Kukla, G. 154
Kuman, K. 68, 100
Kutzbach, J. 2, 148, 161

Labeyrie, L. D. 146, 157
Labracherie, M. 146, 147
La Marche, V. C. 38, 39, 138
Lamb, H. H. 21
Lambrechts, J. J. N. 35
Lancaster, I. N. 9, 30, 31, 32, 48, 49, 50, 52, 54, 61, 62, 63, 64, 65, 66, 69, 71, 72, 154, 155, 156, 164, 165, 166, 177, 183, 185
Lancaster, J. 49
Leakey, L. S. B. 45
Leavitt, S. W. 39
Lee, R. B. 7, 46
Lewis, A. D. 63
Linton, D. L. 34, 123
Long, A. 39
Lorius, C. 157
Lotto, R. 154
Louw, G. N. 49

Mabbutt, J. A. 5, 6, 54, 117
Maggs, T. 38
Mallick, D. I. J. 30, 62, 66
Mametse, M. N. 15
Manabe, S. 19, 20
Manhire, A. H. 120, 121, 159
Marais, M. 120
Marker, M. E. 31, 33, 54, 77, 96
Martin, A. K. 28, 166
Martin, A. R. H. 9, 37, 111, 112, 120, 121, 122, 136, 137, 138, 147, 159, 160, 167, 172, 177
Martin, H. 5, 50, 54
Mason, R. J. 35, 165
Mazor, E. 36
Maud, R. R. 34, 87, 88, 89
Maufe, H. B. 5
McCarthy, M. J. 88
McGee, O. S. 90
McNaughton, J. 38
Meester, J. 41
Merlivat, L. 157
Metcalfe, S. E. 10, 167, 174, 175
Miller, D. E. 120, 121, 159
Minter, W. E. L. 60
Mitchell, J, M. 2, 148, 161
Morley, J. J. 23, 27, 166
Mouton, P. le F. N. 42
Muller, D. 54

Netterberg, F. 9, 35
Newell, R. E. 23, 64, 166
Nicholson, S. E. 10, 24, 25, 158, 172, 173, 174
Nilsson, E. 4

Ollier, C. D. 54
Opdyke, N. D. 10
Opperman, H. 103, 159, 178
Orme, A. R. 88, 89
Owens, J. P. 34

Parkington, J. 120, 121, 129, 130, 159, 175, 177
Partridge, T. C. 9, 24, 31, 32, 33, 57, 77, 85, 97, 105, 152, 153, 155, 158, 159, 164, 166, 181
Peabody, F. E. 76
Penck, A. 5
Pether, J. 113
Petrov, M. P. 30
Pichon, J. J. 146, 157
Pierce, S. M. 168, 169
Pittock, A. B. 10, 21, 24, 167, 174, 175
Pourchet, M. 157
Poynton, R. J. 90, 169
Prell, W. L. 27, 166
Price-Williams, D. 86, 87, 92, 158, 160, 167, 173
Prior, J. 38, 92, 158, 160, 167, 173, 181

Richardson, J. L. 7
Roberts, C. N. 10, 164, 174, 175
Rogers, A. W. 3, 5, 115
Rogers, J. 34, 50, 109, 111, 112, 114, 118, 119, 121
Rognon, P. 50
Ruddiman, W. F. 167
Ruddock, A. 111
Rust, U. 31, 32, 54, 55, 69, 184

SACS (South African Committee for Stratigraphy) 60, 87, 88
Salinger, M. J. 10, 24, 27, 146, 157, 164, 167, 174, 175
Sampson, C. G. 107, 160, 164, 180
Sandelowsky, B. H. 53
Sarnthein, M. 20, 27, 63
Savin, S. M. 2, 148, 161
Schalke, H. J. W. G. 135, 138
Scholtz, A. 9, 37, 38, 40, 41, 123, 131, 132, 133, 137, 138, 146, 147, 153, 154, 156, 157, 159, 160, 164, 166, 167, 168, 169, 170, 172, 173, 175, 177, 178
Schulze, B. R. 12, 14, 15, 16, 90
Schumm, S. A. 31
Schwarz, U. 119
Schweitzer, F. 32, 119, 127, 128, 178
Scott, L. 37, 64, 71, 89, 90, 91, 92, 98, 99, 101, 103, 107, 136, 138, 154, 155, 156, 157, 159, 160, 164, 166, 167, 170, 171, 172, 177, 179, 180, 181
Sealy, J. 43
Seely, M. K. 48, 49, 50, 53, 54
Selby, M. J. 50, 53
Shackleton, N. J. 1, 10, 42, 115, 119, 125, 126, 127, 146, 151, 152, 154, 157, 165
Shackley, M., 51, 52, 53
Shaw, P. 67, 68, 78, 79, 156
Siesser, W. 50
Singer, R. 9, 115, 125, 146, 148, 165, 177
Skinner, A. C. 30, 62, 66
Smuts, J. C. 4, 5, 45
Sohnge, P. G. 5, 31, 105
Solomon, J. D. 4, 45
Sparrow, G. W. A. 9, 96, 97
Sprigg, R. C. 30
Street, F. A. 7, 32, 172, 184
Street-Perrott, F. A. 10, 32, 167, 174, 175
Stuckenberg, B. R. 41, 84
Stuckenrath, R. 24, 31, 32, 33, 76, 77, 98, 105, 107, 155, 157, 158, 160, 165, 166, 167, 179, 180, 182
Summerfield, M. A. 35, 60
Summers, R. F. 45
Szabo, B. J. 34, 105

Talbot, M. R. 30, 63, 64
Taljaard, J. J. 13, 14
Talma, A. S. 24, 33, 36, 62, 64, 85, 86, 97, 142, 143, 153, 155, 158, 159, 160, 164, 166, 167, 178, 181
Tankard, A. J. 32, 33, 34, 50, 60, 113, 114, 115, 117, 118, 119, 121, 123, 127, 128

Tauxe, L. 1
Teller, J. T. 32, 52, 53, 154, 155, 164, 185
Thackeray, A. I. 115, 116, 125, 126, 127, 146, 148, 152, 165
Thackeray, J. F. 9, 36, 37, 38, 40, 41, 59, 74, 115, 116, 123, 125, 126, 127, 131, 132, 134, 136, 145, 146, 147, 148, 152, 153, 154, 156, 157, 158, 159, 164, 165, 166, 167, 168, 169, 172, 177
Thomas, D. S. G. 30, 62, 63, 65, 66, 106
Tissot, C. 37, 50
Tomlinson, R. W. 92
Turon, J. L. 1
Tusenius, M. L. 38, 103, 104, 115, 116, 125, 126, 127, 146, 148, 152, 157, 159, 160, 165, 167, 173, 179
Tyson, P. D. 10, 12, 14, 15, 16, 17, 18, 24, 27, 38, 65, 156, 174

Van Daalen, J. C. 137
Van der Merwe, N. J. 43
Van Riet Lowe, C. 4, 5, 31, 45, 97, 100, 105
Van Rooyen, T. H. 30
Van Straten, D. J. 60
Van Zinderen Bakker, E. M. 9, 10, 22, 23, 24, 36, 50, 66, 73, 74, 75, 76, 79, 93, 98, 99, 100, 125, 130, 136, 137, 164, 166, 181, 182
Verhagen, B. Th. 39, 93
Verstappen, H. T. 66
Vincent, E. 93, 147, 166
Visser, D. J. L. 5, 31, 97, 100, 105
Visser, E. 53, 54, 118, 119, 120, 156, 185
Vogel, J. C. 9, 10, 24, 33, 36, 37, 38, 40, 41, 42, 43, 44, 45, 53, 54, 55, 56, 64, 72, 73, 74, 75, 76, 77, 79, 80, 81, 82, 85, 86, 91, 97, 98, 102, 107, 118, 119, 120, 123, 131, 132, 138, 141, 142, 143, 146, 147, 152, 153, 154, 155, 156, 157, 158, 159, 160, 164, 165, 166, 167, 168, 169, 170, 172, 175, 177, 178, 180, 181, 182, 185

Voigt, E. A. 127
Volman, T. P. 88, 115, 116, 117
Von la Chevallerie, G. 44, 45, 141, 142

Wadley, L. 93
Ward, J. D. 31, 35, 48, 50, 51, 52, 54, 55
Warren, A. 30, 63
Washbourn-Kamau, C. 7
Washington, W. M. 20
Wasson, R. J. 27, 30, 63, 64, 69
Watson, A. 86, 87
Watts, N. L. 60
Wayland, E. J. 4, 5, 61
Wellington, J. H. 106
Wendt, W. E. 55, 56
Werger, M. J. A. 98
Whittington, G. 96
Wienecke, F. 31, 54, 55
Wilkinson, J. 24, 27, 33, 96, 156, 174
Williams, D. F. 166
Williams, J. 20, 27
Willis, J. P. 111
Wilson, M. L. 178
Wright, E. P. 60
Wymer, J. 115, 125, 146, 148, 165, 177

Yaalon, D. H. 35, 50, 88
Yates, R. J. 120, 121, 159
Young, A. W. 177

Zeuner, F. 4
Zilch, A. 31, 32, 105, 107, 157, 179

Subject index

Acheulian 86, 97, 105, 116
aeolian activity, sands, sediments 30, 51, 52, 53, 55, 60, 66, 68, 69, 72, 73, 88, 106, 110–13, 115, 117, 121, 123, 124, 127, 129
Africa
  Central 41, 172, 173
  East 4, 6, 7, 10, 41, 43, 105, 116, 140, 158, 172, 173
  North 10, 15, 24, 173
  North-east 172, 173
  Southern, *see* Angola, Botswana, Cape Province, Lesotho, Mozambique, Namibia, Natal, Orange Free State, Swaziland, Transkei, Transvaal, Zimbabwe, Zululand
  West 114, 119
Agulhas current 12, 28
Aha Hills 68
afromontane forest 136, 137, 138, 170
alcelaphines, *see* mammals
Alexandersfontein Pan 95, 107, 152, 155, 157, 159, 160, 166, 179
Alexandria dune field 111, 121
Algoa Bay 110
Aliwal North sping 37, 95, 99–102, 107, 157, 158, 167, 179
Aliwal Shoal 88
Allerod 99, 166
alluvial deposits, sediments 30, 31, 68, 86, 105, 122, 160, 163, 164, 177
Amanzi spring 109
Aminuis spring 61, 72, 183
Angola 48, 60, 62, 66, 67
Angolan Current 114
Antarctic, Antarctica 2, 9, 10, 21, 23, 42, 48, 157
Apollo 11 Cave 55, 56, 156, 175
aquifers 29, 36, 72, 77, 141, 142, 148, 153, 155, 164, 178
archaeological
  proxy evidence 23, 44, 58, 93, 144, 145, 156
  visibility 45, 46, 58, 172
Arctic 21
Asia 30
astronomical theory 7, 9, 10
Atlantic ocean, coast 12, 48, 54, 94, 109
Atlantic wave 17
Atlas Region 172, 173
atmospheric circulation patterns 10, 12, 13, 16, 17, 21, 23, 24, 25, 30, 64, 65, 117, 165
Auob River 61, 70, 183
Australopithecus, Australopithecus africanus 41, 76
Australia 15, 24, 27, 30, 35, 63, 64, 167, 172
austro-alpine grassland 100

Badzulu gravels 86
Basutolian Ecozone 94–108, 155, 157–60, 167, 179, 180
Beacon Island soil 112, 159
Benguela current, upwelling 12, 48, 50, 114, 130
Berea Formation 87, 88
Berg River 113, 114, 118
Bethlehem 100
Bietou River gravels 122
Blinkklipkop Cave 75
block streams 123, 175
Blue Pool tufa 77, 155
Bluff Sandstone Formation 88
Bok Bay shoreline 117
Bolling 166
Bonawe Cave 38, 103–5, 109, 160, 179
bone collagen 42–5, 56
Boomplaas Cave 9, 38, 83, 109, 130–5, 145–7, 152–4, 156–60, 165, 166–70, 175, 177
Border Cave 10, 33, 80–4, 94, 152, 153, 160, 165, 181
Boteti River 66, 67

Botswana 13, 31, 35, 60, 63, 66, 79, 172
Bovu colluvium 86, 87
Breek Been Kolk Pan 58, 154, 155, 178
Brenton Soil 112, 124, 126, 128
browsers 40, 94, 125, 126, 130, 134, 139, 145–7, 151, 152
Buffelskloof Rock Shelter 38, 84, 85, 109, 133, 169, 170, 178
Bushmanland 46, 56–9, 154
Bushman Rock Shelter 80, 84, 85, 153
Bushmans Fountain spring 95, 107, 180
bushveld 79, 83, 89, 90, 91, 92, 153, 171
Byneskranskop Cave 109, 130, 139, 140, 158, 160, 167, 168, 178

calcified reeds 51, 52, 53, 185
calcium carbonate 77, 86–8, 115, 123
calcrete 9, 34, 50, 52, 66, 68, 69, 73, 105, 107, 111, 113, 117, 123
Cango Caves 36, 109, 131, 141–4, 147, 153, 155, 156, 158, 159, 167, 169
Canis mesomelas, black-backed jackal 40, 76, 140
Cape Agulhas 113
Cape Deseada 114
Cape Ecozone 23, 109–47, 151, 152, 154, 156, 158, 167–9, 177, 178
Cape Flats 13, 109, 135, 136
Cape Fold Mountains 13, 16, 23, 57, 59, 109
Cape Hangklip 109, 138, 177
Cape Province 13, 16, 23, 30
  eastern Cape 79, 86, 97, 170
  northern Cape 13, 46, 65, 106, 155
  north-eastern Cape 38, 94, 98
  north-western Cape 156, 166, 175
  southern Cape 3, 9, 10, 13, 16, 23, 28, 32, 34, 35, 37, 38, 40, 41, 42, 83, 109–48, 153, 154, 159, 160, 166–70, 175
  south-western Cape 13, 16, 17, 23, 32, 34, 35, 38, 40, 42, 109–47, 155, 159, 160, 166
  western Cape 16, 109–47, 155, 160, 166, 169, 175
Cape St Francis 109, 119, 120
Cape St Lucia 80
carbon dioxide 1, 42
carbon isotopes (C_3, C_4, ^{13}C) 39, 42, 43, 44, 55, 56, 98, 143, 144, 153, 155, 169, 170, 175, 180
carnivores, carnivore lairs 40, 43, 117, 134, 139, 140, 141, 164
Cave of Hearths 84, 165
cave sediments, deposits 9, 32, 33, 34, 37, 40, 74, 79, 80–6, 93, 102–3, 115–18, 123–35, 152, 153, 155, 163, 164, 169, 177, 180, 181, 182
cave sinter 26, 69, 86, 154, 184, 185
Cedarberg 138
charcoal analysis 37–9, 92, 94, 103–5, 131, 132, 133, 138, 145–7, 153–60, 163, 166, 167, 169, 170, 175, 177–9, 181
circulation patterns, *see* atmospheric
CLIMAP 19, 20
climatic models 9, 10, 12, 19, 21–8, 174–6
colluvial deposits 30, 58, 68, 77, 86, 106, 122–3, 163, 164
Colwinton Cave 38, 103–4, 109, 158, 160, 179
Conception reed beds 49, 51, 53, 156, 185
Congo River 24
continental drift 5
continental shelf 118, 119
coprolites 107
coral 89
Cornelia pollen site 95, 101, 155, 160, 180
Craigrossie pollen site 95, 101, 155, 157, 160, 180
Cretaceous 60, 122, 124
Crocidura flavescens 83, 134
Cuando River 60
Cuito River 60

De Hangen Cave 109
Damaraland 49
Deelpan hyaena coprolites 95, 107
deep-sea cores 4, 9, 10, 11, 34, 37, 50, 94, 148, 154, 157, 166
dendrochonology 38, 138
dendroclimatology 38

deuterium, D/H ratio 36, 39
diatoms, diatomite 53, 56, 121, 137, 159
Die Kelders Cave 109, 117, 123, 127–8, 139, 140, 145, 152, 153, 160, 166
Diepkloof Cave 109, 129, 153
Dobe Valley 68
Drakensberg 23, 33, 79, 94, 95, 97
Drotsky's Cave 33, 69, 79, 156, 160
Dryas, Older and Younger 166
Duinefontein site 109, 117, 118
dunes 30, 31, 34, 48, 51–4, 62–5, 68–71, 78, 87, 88, 110–13, 115, 116–18, 121, 122, 124, 128, 147, 154, 155, 159, 165
Durban 80
duricrusts 34, 35

Earlier Stone Age 51, 53, 66, 73, 86, 97, 116
eboulis sec 81, 82, 131, 175
Echo Cave 85
Elands Bay Cave, EBC 109, 114, 119, 120, 123, 128–30, 139, 155, 157, 159, 160, 166, 167, 175, 177
Elandsfontein fossil site 109, 116, 118, 140
El Nino, *see* Southern Oscillation
Emambeni Cave 84
Equids 39, 103, 126, 134, 145, 146, 158
Equus Cave 61, 76, 140, 159, 181
Equus sp. teeth 55–6
Estcourt soil 87
estuarine deposits 88, 89
Ethiopia 4, 172
Etosha Pan 32, 69, 158, 184
Europe 4, 35, 135, 148, 166
evaporation, evapotranspiration 4, 16, 19, 27, 32, 36, 60, 63, 64, 70, 76, 77, 115, 165, 166
evaporites 114, 115

fauna, faunal remains 3, 9, 29, 32, 39, 40, 42, 47, 117, 124, 130, 133, 134, 138–41, 145, 153, 163, 164
Faunal Spans 39
ferricrete, ferruginous sands 34, 86, 117, 118, 124, 128, 155
fish 124
Flandrian transgression 111
floral remains 9, 164
Florisbad spring 95, 99–100
Florisian Land Mammal Age 39
fluvial deposits 31, 34, 50, 52, 54, 55, 58, 69, 89, 122
fossil bones, *see* faunal remains
frogs 41
frost, frost weathering 33, 34, 74, 75, 77, 78, 81, 82, 89, 96–8, 102, 108, 123–30, 138, 153, 154, 180
fynbos 40, 101, 110, 136–8, 168, 170

Gaap Escarpment, Gaap Plateau 33, 61, 76–8, 98, 152–5, 157, 160, 166, 167, 182
Gamblian Pluvial 5
gastropod shells 51, 52
geomorphological studies 9
General Circulation Models, GCMs 20
Gi Pan site 61, 68, 78, 184
glaciers, glacial chronology 2, 4, 5, 6
global circulation models 18, 19, 20
Gobabeb reed bed deposits 51, 52, 54, 185
Gobabis 61, 156, 183
Good Hope Shelter 103
Gorrokop tufa 77
Gouritz River 121
grassveld 23, 40, 44, 68, 73, 74, 76, 83, 84, 90, 91, 94, 99, 100–2, 125, 133, 137, 147
grazers 40, 94, 124, 126, 130, 134, 139, 145–7, 151
Great Fish River 79
Griqualand
  East 58, 95, 103–5, 157, 158, 160, 167
  West 46, 58
Groenvlei pollen site 37, 109, 120, 121, 136–7, 147, 159, 167, 177
Grootvloer Pan 58
groundwater 32, 33, 35, 36, 51, 53, 69, 70, 72, 73, 78, 85, 100, 155, 141–4
Guinean Ecozone 172

Gunz 5
Gwayang pollen site 137
gypcrete 34

Haaskraal Pan 57, 155
Hangklip, *see* Cape Hangklip
Harts River 76, 95
Heuningneskrans Cave 80, 84, 85, 154
Herolds Bay Cave 109, 115, 145
Hoedjies Punt fossil site 140
Hoezar Oost Pan 58, 155, 178
Holocene
  general 1, 2, 4, 7, 41, 54–6, 58, 68, 69, 73, 75–7, 79, 83–7, 89, 92, 94, 98, 100, 101, 103, 106, 107, 112, 113, 121, 122, 124, 126–8, 130, 132, 134, 138, 139, 143, 145, 147, 154, 157–60, 167–9, 173
  Early 7, 68, 79, 84, 87, 92, 93, 94, 103, 107, 108, 111, 112, 113, 129, 130, 139, 147, 158–60, 167–70, 172, 175
  Middle 28, 66, 79, 84, 93, 94, 103, 106, 112, 113, 121, 128, 130, 137, 139, 147, 156, 159, 160, 167–9, 172
  Late 45, 46, 58, 68, 71, 79, 84, 85, 92, 93, 94, 98, 102, 103, 107, 108, 121, 122, 130, 137, 139, 140, 146, 158, 159, 160, 169–71
Homeo silts 51
Homo sapiens sapiens 80, 125
Hottentots Holland mountains 123
Howiesons Poort industry 81, 102, 126, 127, 129, 132, 145, 146, 151, 152, 165
Humansdorp 111
human population distribution 46, 161, 171–3, 175
Huon Peninsula sea levels 119, 127, 152, 157, 165
hydrogen 36

India 30, 63
Indian Ocean 12, 13, 16, 17, 24, 27, 65, 93, 94, 154
Intertropical Convergence Zone, ITCZ 12, 13, 17, 21, 23, 24, 174
interpluvials 5, 6, 8, 105
Inyanga Highlands pollen site 92
ionium dates 115, 125
Iron Age 38, 45, 80, 81
isotope studies 29, 33, 35, 72, 97, 108, 141, 148, 163, 169, 178, 180, 181; *see also* carbon, nitrogen, oxygen

Jubilee Shelter 80, 93

Kaffirkop charcoals 138
Kageran Pluvial 5
Kalahari 3, 14, 15, 16, 23, 24, 30, 31, 35, 60–73, 78, 90, 154, 155, 156, 165, 166
Kalaharian Ecozone 60–79, 154, 155, 157–60, 164, 165, 167, 175, 181–4
Kalahari sands 60, 65, 70
Kalkbank site 84
Kamasian Pluvial 5
Kangkara Cave 109, 130
Kanjeran Pluvial 5
Kannikwa Pan 52, 155, 184
Karkkloof tree rings 80
Karoo 16, 46, 56–8, 99, 155
Karoo–Namaqualian Ecozone 48–59, 154, 155, 165, 178
Karpfenkliff conglomerates 51
karst landforms, waters 33, 68, 77
Kathu Pan 61, 73–5, 78, 155, 157, 159, 167, 182
Keurbooms Formation 122
Khesekwe Pan 71
Khommabes Pan 49, 51–3, 185
Kimberley 95
Klasies River Mouth Caves 10, 42, 109, 115, 116, 123, 125–7, 139, 145, 146, 148, 152, 153, 165, 177
Klein Awas Pan 61, 73, 183
Klipfontein Pan 107
Knysna 109, 111, 138
Knysna forest 38, 110, 111
Koes Pan 61, 70, 71
Koichab Pan 49, 52, 185
Kolpochoerus paiceae 88, 116
Komati River 80, 86
Koopan Suid Pan 61, 71

Kruger Cave 39
Kuiseb River 49, 51–55
Kunene River 54
!Kung San 172
Kuruman 70, 74, 95
KwaZulu 80; *see also* Zululand
Kwihabe Hills 61, 68, 69, 78, 184

lacustrine carbonates, deposits, sediments 31, 32, 50, 51, 52, 66, 68, 69, 106, 107, 164, 179, 184, 185
Ladybrand, *see* Rose Cottage Cave 102
lagoonal deposits 88, 114
lakes and palaeolakes 30, 31, 32, 36, 37, 53, 57, 58, 65–8, 72, 107, 136
lake levels 3, 4, 7, 10, 11, 31, 32, 58, 66, 67–9, 78, 107, 108, 154, 155, 158, 165, 172
Lamberts Bay 109
Land Mammal Ages 39
Landsat imagery 29
Langebaan Lagoon sea levels 113, 121, 159
Langebaan Limestone Member 112
Last Glacial 2, 10, 24, 26, 28, 33, 41, 42, 46, 82, 84, 85, 94, 97, 98, 111, 112, 117–20, 123, 125–8, 139, 143, 145, 147, 151, 156, 165, 171, 172, 175
Last Glacial Maximum 10, 18, 19, 20, 21, 23, 24, 27, 38, 55, 56, 58, 63, 68, 74, 78, 82, 84, 88, 89, 94, 97, 102, 111, 118, 119, 121, 125, 128, 130–2, 134, 139, 141–6, 155, 156, 164–7, 169, 170, 174, 175
Last Interglacial 2, 3, 10, 33, 42, 46, 82, 84, 88, 94, 105, 111–18, 121, 122, 124–7, 136, 139, 145, 151, 152, 165, 172
Later Stone Age, LSA 10, 46, 51, 58, 67, 68, 85, 86, 101, 102, 125, 126, 128, 144, 157, 172
Lebombo Mountains 80
Lesotho 16, 33, 55, 94–8, 101, 154
Limpopo River 41, 92
Little Caledon River, Valley 100–2
Little Karoo 110
lizard distribution 42
Lizembe gravels 86
Loerie peat deposit 109, 122, 177
Luderitz 49, 119, 120
lunette dunes 70–3, 106

Mababe basin 67
Madibamatso River 98
Makalian Wet Phase 5
Makgadikgadi Pan 32, 60–2, 66, 67, 70, 78, 79, 154, 160, 184
Malagasy 27
mammals
  larger 39, 41, 42, 74, 79, 84, 88, 94, 100, 110, 124, 125, 130, 133–5, 138–41, 145–7, 151–60, 169, 177
  alcelaphines 84, 103, 126, 134, 145, 146, 158
  extinct 39, 100
  molerat, Cape molerat 128–30, 139, 140, 155, 157
marine sediments 109
marine shell 42, 115, 116, 119, 125–7, 129, 130, 146, 147, 151, 152; *see also* molluscs
Matopos Hills 93
Maun 61
Mauretania 119
Mazelsfontein tufa 77
Mekhaaba pollen site 98
Melikane Cave 55, 95, 97, 98, 153, 154
Melkbosstrand palaeosol 113, 155
Melkhoutboom Cave 109, 158
Meob reed beds 49, 53, 156, 185
micromammals 38, 40–42, 55, 74, 75, 81, 82, 84, 94, 125, 126, 128–30, 134–6, 145–7, 152–60, 166–9, 177, 178, 181, 182
Middle Stone Age, MSA 3, 10, 52, 53, 58, 68, 80–3, 86, 97, 102, 105, 107, 115–18, 124–30, 132, 139, 140, 144, 148, 151, 152, 165, 172
Mier country 70
Milankovitch 7, 9
Mindel 5
Miocene 50, 76, 138
Miombo Region 172, 173
Mlaula Cave 84
Modder River 100
Mogacha fossil valley 31

molluscs 26, 42, 52, 53, 58, 67, 88, 89, 111, 113, 114, 117, 119, 121, 154, 178, 183–5
Molopo River 61, 62, 70, 73, 183
Moreletta River pollen site 80, 92, 159, 160, 181
Mozambique 15
Mozambique channel 93, 147
Mozambique current 12
Mphunga colluvium 86, 87

Nakuran Wet Phase 5
Namib Dessert 13, 14, 31, 41, 48–56, 154–6, 166, 184, 185
Namib Sand Sea 32, 49, 51, 52, 56, 64
Namib IV fossil site 49, 51, 52
Namibia 5, 13, 15, 23, 24, 31, 35, 36, 55, 60, 63, 70, 79, 116, 119
Namutoni spring deposits 69
Narabeb calcareous deposits 49, 51, 53, 54, 185
Natal 16, 30, 34, 38, 41, 79, 87, 89, 95, 97, 99, 103, 113
Natal Basin 93
Nelson Bay Cave 109, 123–8, 139, 146, 147, 153, 155, 158, 165, 177
New Guinea 119
New Zealand 27, 167
Ngami, Lake Ngami 61, 67, 79
Ngamiland 68
nitrogen isotopes 42, 44, 45, 141
nitrates 36, 72
Nooitgedacht Cave 109
Noordhoek pollen site 138
Norga peat 109, 137, 159, 177
North America 2, 35
Nossob River 61, 70

ocean cores, *see* deep sea cores
ocean temperature, *see* sea surface temperatures
Okavango Delta 60–3, 66–8, 78, 156, 159
Olifants River 5, 6, 48
Oligocene 138
Omaramba fossil valley 31
Onink Cave 84
Orange Free State 23, 30, 79, 94, 95, 98, 100, 102, 106
Orange River 54, 60–3, 70, 95, 113, 119, 120, 139
ostracods 115
Otjimaruru Pan 61, 72, 156, 183
Outeniqua Mountains 122
owls 40, 41, 82, 164
oxygen isotopes (^{16}O, ^{18}O) 9, 36, 39, 42, 85, 86, 125–7, 131, 141–4, 146, 151, 152, 154, 158, 159, 165, 181
oxygen isotope stages, chronology 1, 9, 10, 11, 37, 80–2, 115, 117, 123, 124, 126, 127, 141, 146, 148, 152, 153, 157, 165

Paardeberg Cave 130
Pacific core 11
palaeolakes 31, 32
palaeosols 34, 35, 50, 58, 77, 107, 111–13, 118, 124, 155, 166
palynology 9, 23, 36, 73, 93, 137, 163; *see also* Pollen
pans and pan deposits 16, 30, 32, 35, 37, 46, 52, 56–8, 60, 66, 67, 68–74, 78, 94, 95, 106–7, 164–6
peat, peaty deposits 36, 37, 52, 67, 73, 88, 95, 99–100, 112, 122, 137, 138, 184
pedogenic calcretes 105
pedogenesis 9, 34, 35, 58, 66, 69, 87, 112, 123, 124
periglacial landforms/deposits 9, 33, 34, 95–8
photosynthesis, photosynthetic pathways 42
phytogeography 168
Pietersburg 80, 85
planktonic foraminifera 93, 147
Pleistocene
  general 1, 2, 5, 6, 7, 33, 34, 37, 46, 50, 61, 76, 80, 87, 96, 109, 111, 112, 148, 170
  Lower/Early 1, 50, 51, 54, 68, 77, 86–8, 92, 105, 115, 117
  Middle 1, 54, 58, 66, 67, 74, 77, 87, 88, 100, 105, 116, 118, 129

Upper/Late/terminal/end 1, 2, 10, 32, 39, 41, 48, 50, 55, 56, 58, 68, 69, 73, 75, 76, 78, 79, 82, 84–8, 92, 94, 97, 100, 103, 105, 107, 111–13, 116, 118, 124, 128, 130, 135, 138, 139, 143–7, 157, 158, 167, 169, 170, 172
Plettenberg Bay 111, 122, 124
Pliocene 1, 2, 76, 138
pluvial scheme, pluvials, interpluvials 4–9, 105, 117, 148
pollen 36, 37, 50, 66, 68, 73–7, 89–92, 94, 95, 98–102, 106, 107, 108, 131, 132, 135–9, 145–7, 155–7, 159, 160, 163, 164, 167, 169, 170, 177, 179, 180–3
Pomongwe Cave 84
Port Durnford fossil site 80, 88
Port Nolloth 52
Present Interglacial 2, 10, 38, 46, 139, 143, 151, 165, 172
Protea Banks 88

radiolaria 93
rainfall/precipitation
general 2–5, 7–10, 12–16, 19, 20, 27, 28, 30, 32, 36, 38–41, 44–7, 55, 62–4, 67, 69, 70, 75–9, 84, 89, 94, 99, 105, 109, 123, 138, 148, 164, 168, 174, 175
drier 13, 20, 24–8, 35, 41, 50, 52, 53, 55, 56, 58, 64–9, 71, 73–5, 77–9, 81–3, 85, 87, 89, 90–3, 94, 98–103, 106, 107, 108, 112–15, 123, 126, 128–30, 132–5, 137, 139, 146, 147, 150, 151–61, 165–70, 172, 174, 175, 177–84
wetter 13, 24–8, 40, 50, 52–6, 58, 59, 66–9, 72–9, 81–3, 85, 87, 89–94, 99, 100–3, 105–8, 112, 113, 117, 123, 124, 126–30, 132–7, 139, 140, 145–7, 150, 151–61, 164–70, 172, 174, 175, 177–84
summer 13, 14, 17, 18, 21, 24, 26, 27, 28, 43, 47, 49, 56, 57, 59, 60–109, 132, 137, 147, 150, 154, 156, 159, 161, 165, 166, 168, 174–6
winter 13, 14, 17, 18, 21, 23, 24, 26–8, 43, 47, 49, 55, 56, 59, 102, 105, 109–47, 151, 152, 154–7, 161, 166, 167, 168, 169, 174–6
all-year 13, 43, 47, 109–47, 151, 154, 159, 161, 166, 169, 170, 175, 176
Ravenscraig Cave 38, 103, 104, 109, 158, 160, 179
Redcliff Cave 153
Rietputs Formation 101, 105
Riet River 95, 105
Rietvlei pollen site 80, 91, 92, 109, 135, 136, 159, 180
Riss 5
river terraces 4, 5, 6, 30, 31, 97, 100–1, 105–6, 122
Riverton Formation 95, 101, 105, 106, 152
Robberg Peninsula palaeosols 111, 112, 124
Rocky Bank 118
Rooidam Middle Stone Age site 95, 105
Rooikrans aeolian deposits 112
Rose Cottage Cave 95, 98, 102–3, 152, 153, 158, 160, 180
Rössing Cave 49, 55, 154, 185

Sahara 30, 50, 63, 172, 173
Sahel 30, 63, 64, 172
Saldanha Bay 113, 116, 117, 120
San Nicolau 48
Sani Pass 95, 96
savannah 43, 62, 63, 74, 75, 76, 83, 89, 91, 94, 170
Savuti River 67
Scot pollen site 80, 92, 159, 181
Sea Harvest fossil site 109, 139, 140
sea levels 3, 9, 34, 42, 87–9, 111–22, 124, 125, 127, 129, 130, 137, 145–8, 152, 157, 159, 165
sea surface temperatures 19, 21, 23, 26, 28, 34, 36, 42, 89, 93, 127, 147, 148, 166
sediment analysis 3, 9, 10, 29, 66, 80, 81, 100, 123–6, 128, 140, 152, 163, 164, 169, 179, 180, 182

Sehonghong Cave 180
shell, shellfish, *see* marine shell, molluscs
Shongololo alluvium 86
Shongweni Cave 84
shorelines 34
silcrete 34, 35
sinter, *see* cave sinter
Siphiso Cave 38, 80, 92, 94, 158, 160, 167, 181
Skurwerug fossil site 116
slope deposits 122–3
snail shells, *see* molluscs, gastropod shells
snakes 41
snow 96
South America 2, 63
South Atlantic Ocean 12, 13, 17, 24, 27, 65, 114
South Indian Ocean 24, 27
Southern Oscillation 17, 18, 24, 27, 28, 174
Sossus Vlei Sand Formation 49, 51
speleothems 29, 33, 35, 36, 55, 69, 78, 85, 86, 115, 125, 131, 141, 143–4, 147, 148, 155, 156, 158, 159, 164, 167, 169, 181
spring activity, deposits 69, 72, 76, 77, 85, 89, 99–100, 102, 106, 107, 158, 170, 179, 180, 183
stalagmite 24, 69, 75, 85, 126, 153, 169
Stampriet Pan 36, 61, 72, 158
Stanford palaeosols 109, 113, 117
St Helena Bay 113
Stillbay 112
St Lucia 89
Stone Age 3, 45, 46, 51, 156; *see also* Earlier, Middle, Later Stone Age
stromatolites 72, 73, 183
subtropical convergence 27
Swakopmund 55
Swakop River 49
Swartberg Mountains 123, 130
Swartklip fossil site 40, 109, 113, 117, 139–41
Swartkops fossil marine molluscs 89, 111
Swaziland 39, 80, 84, 86, 92, 158, 160, 167

Taung fossil site 76–8
tectonic activity, tectonism 34, 54, 67, 68, 113, 128, 169
temperature
  general 2, 3, 4, 8, 10, 15, 16, 19, 20, 21, 23, 24, 36, 38–41, 43, 44, 46, 64, 67, 83, 84, 94, 136, 138, 139, 141, 148, 162, 164, 168
  cooler 6, 20, 23, 24, 27, 33, 34, 40, 41, 43, 53, 56, 64, 67, 74–8, 80–5, 89, 90, 92–5, 97–103, 105, 107, 108, 117, 124–8, 131–6, 139, 140–7, 149, 152–60, 161, 164–6, 169, 170, 172, 174, 175, 177–84
  warmer 27, 43, 64, 66, 71, 73, 74, 79, 81, 82, 85, 87, 89, 90, 92, 93, 94, 99, 103, 107, 108, 114, 115, 117, 121, 124, 125, 126, 128, 132–7, 139, 146, 147, 149, 152, 158–61, 165, 167, 170, 174, 175, 177–84
Tertiary 50, 51, 77, 105, 109, 122
Thamakalane fault 67
Thar Desert 63
Theuniskloof pollen site 80, 92
Tortoise Cave 139
Transkei 79
Transvaal 13, 24, 33, 37–9, 41, 63, 77, 79–94, 140, 154–6, 158, 164, 168
Transvaalian Ecozone 79–94, 152, 154, 158, 159, 160, 165, 167, 170, 180, 181
travertine 33, 76, 84
tree rings 38–9, 92, 138, 164
Tsams Pan 54
Tsauchab Pan 31, 49, 54
tsetse fly distribution 45
Tsitsikamma Coast 111, 122
Tsondab Pan 31, 49, 51, 53, 54, 185
tufa 33, 68, 72, 76–8, 152–4, 182, 183
Tugela River 80, 92

Uitenhage aquifer 36, 86, 109, 141–3, 147, 153, 155, 159, 164, 167, 168
Uganda 4
Upington 61, 63, 70
uranium disequilibrium dating 116, 152

uranium/thorium dating 53, 77, 105, 152, 153
Urwi Pan 61, 71, 72, 156, 183

Vaal River 5, 31, 77, 95, 101, 105–6, 152
Van Riebeeckstrand palaeosol 112
vegetation 7, 23, 30, 37, 38, 40, 47, 49, 50, 53, 57, 62, 63, 66, 68, 73, 74, 76–9, 83, 84, 87–92, 94, 96, 98–103, 106, 107, 112, 113, 121, 125, 128, 132, 134, 136, 138, 139, 143, 145–7, 151, 153, 156–8, 168–71
Verlorenvlei deposits 109, 113, 114, 121, 129
Victoria Falls 61
vleis 36, 37
Voigtspost Pan 107, 160, 180
Vrendenburg shell beds 113

Walker circulation 17, 18, 27, 28
Walvis Bay 49
Walvis Ridge 50
Wedberg Terrace 77
Welgevonden alluvial and colluvial deposits 102, 106
West Pan 49, 51, 52
Wilderness dunes, lakes 109, 111, 112, 121, 122
Wilton Cave 109, 158
wind 16, 19, 21, 23, 24, 26, 27, 28, 30, 34, 37, 41, 62–6, 70, 71, 88, 112, 118, 129, 132, 153, 155, 165, 174
Winn alluvium 86, 87
Wolkberg Cave 80, 85, 86, 153, 155, 158, 159, 167, 181
Wonderkrater pollen 37, 80, 89–91, 94, 154, 156, 158, 159, 160, 167, 168, 170, 171, 181
Wonderwerk Cave 61, 74–6, 78, 95, 98, 153, 159, 167, 182
Wurm 5

Xangwa Pan 68, 184
X-ray diffraction 29
xylem analysis 38, 132, 133, 154, 166, 175

Yebo alluvium 86
Ystervarkpunt 121

Zaire 119
Zaire Air Boundary 13
Zambezi River 1, 5, 67
Zambia 45, 60, 62, 66
zebra teeth 55, 98, 155
Zimbabwe 5, 15, 31, 41, 45, 60, 63, 66, 79, 84, 92, 93, 153
Zoetvlei River 58
Zululand 87, 89; *see also* KwaZulu